PRAISE FOR KELLY BRENNER'S FIRST BOOK, *NATURE OBSCURA*

"Readers will be both mesmerized by the surprising poetry of Brenner's prose and excited to follow her example and discover nature's marvels in their own surroundings."

—*Booklist*

"Brenner brings an infectious curiosity to urban nature. . . . With *Nature Obscura*, readers need not venture far to discover a natural world teeming with life."

—*Shelf Awareness*

"This unofficial guide to finding wildlife in unexpected areas is wide ranging and companionable. Also included are tips and resources for budding urban naturalists."

—*Library Journal*

". . . embraces the mystery even as it searches for answers . . . it's a relief to be transported via Brenner's calm, elegant prose to the worlds of tardigrades and snails, moths and molds."

—*The Seattle Times*

"Drawing upon her astonishing fine ability to notice, and find both fascination and enjoyment in, things that are all-too-easily missed in the daily hurly-burly of the modern world—mosses, ferns, crows, muskrats, slime molds (especially slime molds)—Ms. Brenner uses her exquisite command of the English language to guide her readers along on a tour of the wonders to be found in the natural world that is not far-off and exotic, but close, easily accessible, and remarkably (for lack of a better word) cozy."

—*The Well-Read Naturalist*

"Impeccably researched and written, Brenner's smart debut is ideal for those with a deep interest in nature."

—*Publishers Weekly*

". . . Brenner's book is a comforting reminder that nature can be spotted just about anywhere."

—*Outside Online*

"Naturalist Kelly Brenner is keen to show that nature is 'within easy access.' "

—*BBC Wildlife* Magazine

"A warm-hearted meditation on the natural wonders that we city dwellers overlook every day, such as the western pondhawk (a dragonfly), licorice ferns, and the charming Anna's hummingbird, her heart beating 1,250 times a minute—and the perhaps less charming ant-decapitating fly and, yes, *Fuligo septima*, a.k.a., dog-vomit slime mold . . . Once you read this book you'll step more carefully, to avoid damaging the myriad tiny worlds that add color and texture to our world."

—Erik Larson, author of *The Splendid and the Vile*

"Gorgeously written and deeply felt, with rigorous research and information at its core."

—Jeff VanderMeer, author of the *Southern Reach* trilogy

"*Nature Obscura* reveals the living city through delightful encounters with natural neighbors we all ought to know. An attentive, fun, and thoroughly engaging guide, Kelly Brenner will help you become a better naturalist every day."

—Robert Michael Pyle, author of *Sky Time in Gray's River*

"With observant eyes and beautiful prose, Kelly Brenner draws us all into the hidden depths of the urban wilderness. Hummingbirds, dragonflies, ferns, and even slime molds come to vibrant life alongside stories of the humans who keep watch over the nature that surrounds us. *Nature Obscura* inspires everyday wonder, adventure, and wisdom about our changing earth."

—Lyanda Lynn Haupt, author of *Mozart's Starling* and *Crow Planet*

The Naturalist at Home

Projects for Discovering the Hidden World Around Us

Kelly Brenner

MOUNTAINEERS
BOOKS

MOUNTAINEERS BOOKS is dedicated
to the exploration, preservation, and enjoyment
of outdoor and wilderness areas.

1001 SW Klickitat Way, Suite 201, Seattle, WA 98134
800-553-4453, www.mountaineersbooks.org

Printed in China
Distributed in the United Kingdom by Cordee, www.cordee.co.uk

26 25 24 23   1 2 3 4 5

Copyeditor: Jennifer Kepler
Design and layout: Jen Grable
All illustrations by the author. Cover illustrations: front: Acorn barnacle (Project 4); curiosity cabinet (Project 20); underwing moth (Project 8); back: lacewing (Project 11); springtail (Project 6)

Library of Congress Cataloging-in-Publication Data is available at https://lccn.loc.gov/2022039437. The ebook record is available at https://lccn.loc.gov/2022039438.

Mountaineers Books titles may be purchased for corporate, educational, or other promotional sales, and our authors are available for a wide range of events. For information on special discounts or booking an author, contact our customer service at 800-553-4453 or mbooks@mountaineersbooks.org.

Printed on FSC®-certified materials

ISBN (paperback): 978-1-68051-573-2
ISBN (ebook): 978-1-68051-574-9

An independent nonprofit publisher since 1960

There is a pleasure in the pathless woods,
There is a rapture on the lonely shore,
There is society where none intrudes,
By the deep Sea, and music in its roar:
I love not Man the less, but Nature more,
From these our interviews, in which I steal
From all I may be, or have been before,
To mingle with the Universe, and feel
What I can ne'er express, yet cannot all conceal.

—Lord Byron, *Childe Harold's Pilgrimage*

Contents

Introduction: Becoming a Naturalist

"A naturalist is lucky in two respects. First, he enjoys every bit of the world about him and has a much more enriched life than someone who is not interested in nature. Second, he can indulge his hobby in any place at any time, for a naturalist will be fascinated to watch nature struggling to exist in the midst of a great city as well as observe its riotous splendor in a tropical rainforest."

—Gerald Durrell, *The Amateur Naturalist*

Even in July the waters of Puget Sound are so cold that if a naturalist wades in wearing sandals, their toes can become so numb they won't notice a broken shell stuck in their foot. I stood stoically, however, in that cold water, finally wise enough to wear insulated chest waders, searching for nudibranchs.

Forget the idyllic image of relaxing on a sunny, sandy beach dotted with palm trees. Instead, I was dockfouling, exploring under a ferry dock where the water is permanently murky from the ferries churning up the sand. Although my feet and legs were mostly insulated, my hands were numb to the point of becoming frozen in place, fully submerged in the water, holding an underwater camera aimed at a blob of white.

Using my aquascope, I had discovered a nudibranch. A frosted dirona (*Dirona albolineata*) nudibranch, to be precise. Their name is appropriate, as these sea slugs look just like frosted glass, translucent and decorated with brilliant white markings along their bodies and cerata—the fingerlike projections on their backs. A very beautiful and elegant slug.

When I glanced to the right, I yelped with excitement to see an orange version of the frosted dirona, a nudibranch I had been hoping to find, called the golden dirona (Dirona pellucida). I quickly took photos of my discovery and pulled my numb hands out of the water to check the images. As I zoomed in, I noted that the golden sea slug was very orange and had the same bright white markings as the frosted dirona. But I felt deflated as I realized that this nudibranch did not have white spots—the signature markings of a golden dirona.

As the waves from the ferry broke against my knees, I studied the strange orange nudibranch on my camera screen. If it wasn't a golden dirona, it had to be a frosted dirona, but how was one individual white as snow, while the other was the color of a citrus fruit? I considered what I knew about nudibranch anatomy. Dironas are part of the Aeolids group, which have cerata projections on their backs. Their cerata are not just for decoration. Many Aeolid nudibranchs consume prey, such as sea anemones, which have stinging cells called nematocysts. The nudibranchs ingest those cells, but here's where an Aeolid nudibranch's anatomy gets curious.

Their unique digestive system flows up through the individual cerata—an adaptation that allows them to take the stolen nematocysts and repurpose them, creating their own stinging defenses. (Only one nudibranch in the world, the blue glaucus [Glaucus atlanticus], found around Australia, can actually sting humans; the rest are harmless unless you're a fish.) Since their digestive system runs throughout their body, was it possible the orange coloration was due to something the nearly transparent frosted dirona nudibranch ate?

I turned around and found the answer spread out before me. In the shadow of the dock, hundreds of bright orange sea pens, an animal that grows in colonies and looks like an old-fashioned quill pen, some growing as tall as my knee, swayed back and forth with the waves. Maybe the orange frosted dirona nudibranch had been dining on the sea pens.

Referencing my field guide back at home, I learned that frosted dironas do indeed consume sea pens. My naturalist's intuition was correct—the sea slugs I saw were all frosted dirona nudibranchs. (For anyone who likes happy endings, the very next day under the same ferry dock I found a real golden dirona, decorated with white spots.)

What Is a Naturalist, and Why Should You Be One?

Many people have offered a variety of opinions about what makes a person a naturalist, both historically and in modern times, but let me offer my own definition first: *A naturalist is someone who observes, studies, documents, or otherwise learns about the natural world.*

And what an incredible world it is. Naturalists are never bored. We're never short on fascination or inspiration because nature is everywhere. The curious naturalist has only to open their senses, no matter where they are, to connect with the natural world. We always have something to entertain us and keep our brains overflowing with questions. Being a naturalist is enriching, healthy, and wondrous. We can practice on our own, with a friend, or with a group of like-minded individuals.

No matter who you are or what your level of experience, I welcome you to join the society of naturalists, which is accessible regardless of your age, location, or education. There are no required degrees, fees, exams, or certifications. It's not a formal society, but you're welcome to join anyway. All you need is an interest in the natural world, a sense of curiosity and wonder, and a desire to explore and learn.

Many of the best naturalists have no formal training. They don't have framed degrees on their walls or fancy

Eubranchus olivaceus

Aeolid nudibranchs

titles. Instead, they have learned what they know by sheer dedication, and they find a great sense of freedom in learning on their own terms. They devote their time and energy to observing nature, reading books about nature, documenting their observations, and generally obsessing over the natural world.

Many of them are now household names. Perhaps the best-known modern naturalist is the legendary Sir David Attenborough, who was drawn to nature as a child, then had a career as a television producer before he turned to making natural history programs, such as *Planet Earth* and *Blue Planet*, that changed television and opened the world of nature up to millions of people.

Rachel Carson was also a student of nature from a young age and wrote about it for most of her life. She penned a trilogy of wonderful books about marine life detailing her personal explorations of the seashore, before writing the book that would make her famous. *Silent Spring* had a massive impact in the 1960s and raised awareness about the previously unrecognized dangers of pesticides.

Gerald Durrell is best known for his humorous books, including the 1956 classic *My Family and Other Animals*, which documented his early life as a child naturalist running around the island of Corfu, catching every animal he could find, much to the chagrin of his long-suffering family, and his later adventures opening his own zoo. Durrell challenged the established midcentury purpose of zoos, believing they should be used for conservation of endangered animals instead of entertainment with little consideration for animal welfare. He helped usher in a new era of zoo conservation.

Countless naturalists have been influenced by Anna Botsford Comstock, whose 1911 book, *The Handbook of Nature Study*, can be found on many naturalists' bookshelves today. Comstock's lifelong work to study and then teach natural history was instrumental in widely expanding the topic to students throughout the United States in the early 1900s. She became the first woman professor at Cornell University, although she still suffered from sexism and was denied full status for twenty years.

Gulielma Lister is less well known today but just as inspiring. Lister was a self-taught British naturalist who worked closely with her father and was influential in the study of slime molds. She illustrated several books about natural history and corresponded with people around the world about slime molds, both sending and receiving many specimens. With her father, she wrote and illustrated a comprehen-

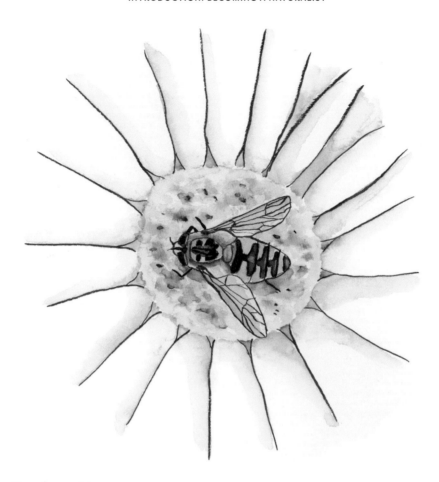

Hover fly on a daisy

sive book on all the known slime molds in the early 1900s, completing new editions after he died.

A hundred years before Comstock and Lister, there was Mary Treat. Fascinated by nature from a young age, Treat had a particular fondness for insects that far outlasted her marriage to a husband who died penniless, leaving her nothing. She began studying natural history in New Jersey in the 1800s. Roaming around the region solo, she collected plants and insects and intensively studied her personal garden patch, known as the "Insect Menagerie." As with most naturalists, her interests didn't lie solely in one area. She was also fascinated by carnivorous plants and corresponded extensively with Charles Darwin on the topic, taking up a dedicated study on them as a result of a particular quandary.

Treat's fascination with the natural world provided her with a needed income as she documented her observations and then eventually wrote *Home Studies in Nature* and several other books and articles. She went on to discover a number of new insect species and document behavior in the wild and with experiments. Treat wasn't afraid to disagree with the establishment and was proved right over the "experts" on multiple occasions. Despite being known as an amateur naturalist, she outshone the experts by popularizing natural history with her work and her writing, which supported her until she died at the age of ninety-three.

Treat and Lister continued a long tradition that can be traced to a pirate naturalist. During his record-setting three times circumnavigating the globe in the 1600s, William Dampier had plenty of time to collect specimens, documenting and analyzing the natural history of the various regions he sailed through while also pioneering new and highly influential methods in the field of marine navigation. His books about his adventures greatly influenced many people, including Charles Darwin and James Cook.

Sadly, the subject of natural history has largely become relegated to the past. In Dampier's day the field encompassed everything from biology to astronomy, geology, anthropology, and even medicine. Today it has a narrower definition tied to the biological part of the natural world, mainly plants and animals in their natural habitat. Where it used to be a prominent part of a classical education, natural history is now largely absent in schools from primary up to higher education, replaced by science. What's the difference? Natural history generally involves observational methods, while science is more experimental. There is a certain amount of overlap of course, and natural history study can use scientific experimentations, but it's not limited by the methods of science. Science experiments often hyperfocus on one small part of the larger natural history—often limited to those organisms humans benefit from or those that harm us—like plucking a piece out of a puzzle to analyze.

Both science and natural history provide invaluable information and data. It's vital to see them as allies, not competitors. That's why it's important not to allow the study of natural history to be resigned to the past, nor overshadowed by science education.

The benefits of studying natural history in the modern world are many. Naturalists have the freedom to wander and focus on whatever catches their attention. Naturalists also enjoy the freedom to think and act outside the rigid rules of science, and to bridge the gaps and tie together scientific fields such as botany, entomology, and zoology. In this, we naturalists can offer a unique perspective. We can observe a habitat, find patterns and connections, learn the language of the landscape, and become well acquainted with the entirety of it. We can change our perspective from looking at the whole of the habitat one day to the minutiae living within it the next. Naturalists have the flexibility to tailor their study to fit their interests and time. We can spend a week studying lichens—or a month, or even a lifetime. We can choose to record everything about a place, or everything about a certain species. Or we can choose to record nothing at all and simply observe. There are no rules when it comes to your personal study.

Jellyfish observed while dockfouling at a marina

How Can This Book Help You Become a Naturalist?

Naturalists can enhance their own personal nature study with aids such as classes, events, and field trips. And books. There are many, many books that are useful for naturalists, from field guides to memoirs written by naturalists. There are textbooks and art books, history books and creative nonfiction. For my personal recommendations, consult the Resources at the end of this book. *The Naturalist at Home* takes its title from a chapter in Gerald Durrell's classic book, *The Amateur Naturalist*, an excellent text for learning all the basics of natural history, which I recommend as an ideal companion book. *The Naturalist at Home* offers something unique: a collection of projects and activities to help you start observing and immediately learning

Naturalist's Toolkit

Here are a few of the most common general naturalist tools, as well as some of materials used most frequently in the projects.

- Hand loupe
- Pen light, preferably waterproof
- Knife
- Shovel
- Containers for collecting: empty tins, paper bags, vials, jars, empty water bottles, plastic food containers
- Small paintbrush (for moving invertebrates)
- Pipette
- Petri dish (or repurposed glass dish)
- Dissecting microscope
- Compound microscope
- Plaster of Paris
- Clay
- Paper
- Artist's fixative spray

about the natural world around you. You can learn a tremendous amount from books, but you will learn infinitely more by getting outside and conducting your own study of natural history.

The projects and activities that follow will help a naturalist, or anyone curious about becoming one, to observe and study everything from invertebrates to mushrooms and mammals. The projects are based on various techniques actively used by naturalists and scientists, some of which have been in practice since the Victorian age or even earlier. Many of these techniques were developed by curious naturalists driven to find a method to learn more about something that had caught their interest. This book will introduce ways to observe many often-overlooked organisms.

I chose projects suitable for anyone of any age or in any location, and of any expertise or experience level. A naturalist who lives in Australia, Iceland, or Seattle can complete these projects. Of course, there are bound to be some locations where individual activities won't work, but for the most part, everyone should have the ability to do most of the projects in this book. I'll guide you through all you need to know, so don't worry if the last time you did these kinds of projects was in elementary school.

You won't need to spend a lot of money on specialized equipment. For the most part, you can make use of supplies you likely already have at home, can easily find at a thrift store, or that are inexpensive. There are almost no specialized tools except perhaps a microscope. If any projects are particularly interesting, consult the Resources for literature to help you dive deeper into any subject. Many projects, such as pond dipping and finding tardigrades, tie directly to my first book, *Nature Obscura: A City's Hidden Natural World.*

While these projects can help you further your personal study of natural history, they can also add to the growth of natural history and science as a whole, by creating data and observations that can be shared with the wider community through community science projects, social media, text messaging, websites, and more. It's the modern version Mary Treat and Charles Darwin sharing observations by post.

Each project in this book provides an overview of the subject and what you may discover. A list of materials outlines what you'll need, and detailed steps guide you through the process. Finally, many projects include summaries of some of the organisms you may encounter.

Projects of similar themes are grouped together, and when done jointly, they can greatly enhance your study. Many projects can also be enhanced when performed repeatedly at different times throughout the year and in different locations. It's important to get out throughout the year, not just when the weather is pleasant. As we like to say in the Pacific Northwest, there is no bad weather, only bad clothing (extreme weather aside).

You can do them in any order, according to whatever interests you most. However, I suggest at least reading through the projects on keeping a nature journal (Project 1), learning the language of nature (Project 19), and creating a curiosity cabinet (Project 20). You are bound to find the tips on general observation of the natural world in those three projects informative as you try the others.

These methods will help naturalists learn about the organisms that live around them. They'll help you change your perspective in thinking only about the larger life-forms and reconsider the life of bacteria or face mites. Most of all, these projects demonstrate how interconnected the natural world is. Anyone with a sense of wonder and curiosity can become a naturalist.

General Naturalist Guidelines

There are entire books about how to look at the world from a naturalist's perspective. There's also a chapter on being a naturalist in my first book, *Nature Obscura*. The biggest trick to beginning to truly see the natural world is to start honing your senses, especially when you're outside. Slow down, and sit for a while, or walk without a destination in mind. Take the time to observe. It takes a long time to "get your eyes in," and one of the most fun aspects of being a naturalist is that you'll never notice everything. There is always something more to discover.

Begin to change your perspective. Instead of focusing on megafauna, look at small organisms or ones that don't move. Challenge the common conception that nature is "out there." If you shift your perspective, you'll discover that nature is up on your roof, in the trees branching over the streets, and growing in the cracks of the pavement. Visit places in the city with hidden nature, like graveyards that host a multitude of lichens or bridges where falcons may nest.

Naturalists should always observe sound ethical practices and be respectful of the natural world. Don't overharvest, disturb wildlife, harm the ecosystem, or otherwise damage the nature you're studying and observing. While you're free to explore your own yard if you have one, if you venture onto private or public land, know the rules. Many parks restrict collecting anything, but often you can get permits to collect things like mushrooms.

You should not only protect and respect nature but also take steps to keep yourself safe. These projects will vary depending on where you live, so please exercise caution if you live in a region with poisonous plants, venomous spiders, or other potentially dangerous wildlife or weather. Always be alert to your surroundings, and if you aren't familiar with the plant or animal you're investigating, avoid close contact with it.

There are countless ways to share what you're learning. You can make your observations public on a blog or website, write posts on social media, or find groups to join online. If you connect with other naturalists, you can exchange information through text messages or emails. If you study an organism or location intensely and share your findings, local nature organizations, clubs, or parks may ask you to give presentations or display your observations.

Community science is a formal way to share your observations and data with organized projects that add to the wider understanding of nature. Two popular examples of websites that collect observational data are iNaturalist and eBird. You can also find very specific projects documenting everything from the distribution of bumble bees to when plants bloom and the timing around migrating dragonflies. Look for and join national or local nature organizations, which frequently conduct community science projects you can participate in. Once you get some experience with the projects, you could even consider running your own on a subject you're working on!

Whether you've already begun your journey as a naturalist or are taking your first steps now, I wish you a lifetime of wonder and joy with the natural world.

Keep a Nature Journal

SEASONS:
All

STUDY TOPICS:
Natural history
documentation,
observation skills

I f you asked someone who keeps nature journals what they'd save first if their house caught fire (other than pets and family), they would likely point to their journals. Why? What makes these books so special?

A nature journal is simply a record of a naturalist's experiences and observations along their journey through natural history. This documentation of meaningful, unique, and special connections with the natural world reflects their understanding and ongoing education. Nature journals record the naturalist's delight in new discoveries and experiences, and reflect the hard work of recording through words and art.

In addition to all of those benefits, nature journals help naturalists gain a deeper understanding of their local—and perhaps

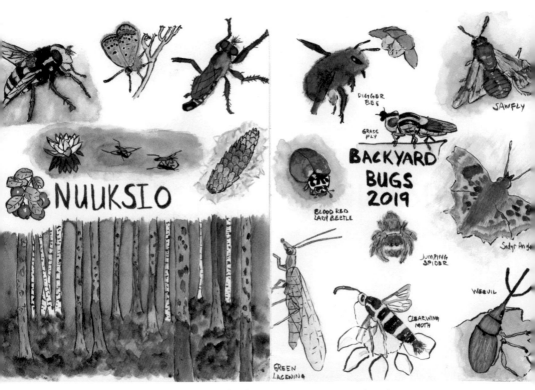

A study of one place at one point in time (left), *and another over a year* (right)

more distant—habitats and what lives there. We all connect with and appreciate a place more when we observe it carefully, and in doing so, we may also discover a desire to care for and protect it.

There is no standard or "right" way to keep a nature journal, and perhaps the freedom of expression is one of the aspects that makes them special enough to cherish and preserve. Your nature journal may include simple observations, notes, and drawings featuring what you observe in the natural world, or you may be inspired to add quotes, song lyrics, poetry (your own or from others), lists, diagrams, and maps. You might glue or tape in pressed plants and feathers, or make bark rubbings or spore prints in it.

I know from experience that artistic skills don't come easily for everyone. Illustration isn't necessary in a nature journal. It's entirely possible to record your observations with words, lyrics, plant pressings, and color. Some naturalists color-code their notes or use watercolors to capture the palette of a habitat instead of drawing a detailed scene. The flexibility of the nature journal format offers each and every naturalist the ability to tailor their approach.

That said, there are huge benefits to trying your hand at drawing, even if you choose not to keep a nature journal long term. The process of illustrating something, whether it's an entire landscape, a single tree, or a tiny insect, helps you see details. When you focus on something with the intensity required to reproduce it on paper, you'll reach a whole new level of observing details like shape, proportions, texture, shadows, and color—all of which make up the whole. A detailed study can also reveal damage and imperfections, which helps uncover the story of that object.

Many artists prefer to draw and add to their journals out in nature, but it's perfectly fine to wait until you're back home and write or illustrate from photos you took. In fact, there are benefits to using photos because they preserve the moment, including the quality of the light, weather conditions, animal movement, and other elements that can change or disappear before you have finished your drawing.

Don't worry about creating professional-quality artwork. Be free and loose. Regardless of how it turns out, every drawing teaches you something, and the more you draw, the better you'll get. Art is a skill just like any other, and one that you can learn. Practice is important, but there are countless resources for learning not only how to create different styles of artwork but also nature art specifically, including field sketching and nature journaling. Consider taking an art class. Study published nature journals. Look for books that teach nature drawing and try the exercises. For a few suggestions to get you started, see the "Nature Journaling" section in Resources.

Beyond improving your drawing and observations skills, keeping a nature journal teaches you patience and improves your ability to focus. When you set aside everything else and pay attention to something

Tip

If you're tracking multiple things, consider keeping a separate journal for each one. Perhaps you'll have a journal devoted to your nudibranch observations, another for home observations, and yet another journal for observations made on a trip.

Journal entries from some nature walks

specific in nature, you have the opportunity to practice meditation and mindfulness. In fact, nature can be a wonderful healer if you consider your nature journal as an opporunity to reflect on your thoughts instead of turning it into a scientific study. How you approach it is entirely up to you. The most important thing is that your nature journal is fun for you, because if it isn't fun, you won't stick with it.

What to document is just as open as how you do it. Perhaps you'll want to record the colors of a place or season, the signs and tracks of animals, specific wildlife behavior, the ways clouds change over a mountain range, or the items you collect on a walk. You can document a particular location or the seasonality of plant life. Maybe you'll study the anatomy of animals and plants, the construction of a spiderweb or wasp nest, or the daily growth of a bud on a tree branch as it transforms into a flower or leaf. Of course, you can also document the many projects in this book. A nature journal is the perfect place to keep track of a slime mold in a moist chamber (Project 15) or the tardigrades found in a patch of moss (Project 6), or to preserve a spiderweb or spore print (Projects 13 and 14, respectively).

What can you do with your nature journal? Your journal may be private, or perhaps you want to share pages on your website or social media. You can also use your observations to

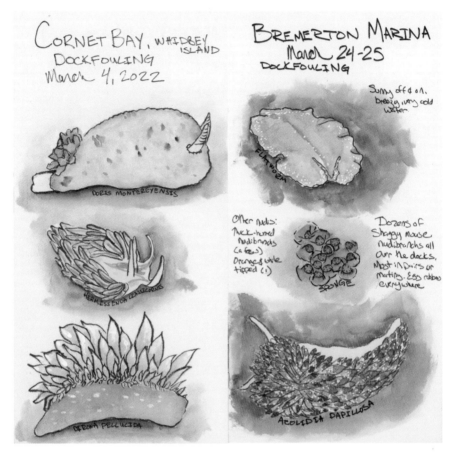

Observations of marine life found while dockfouling

contribute to community science projects. (See the "Community Science Projects" section in Resources for a list of projects.)

Materials

Journal: With countless options, choosing a journal can be overwhelming. There is no right type of journal; it's up to you and how you plan to use it. If you're more likely to write than draw, then a lined or grid-format journal may be most useful. If you work mainly in watercolor, a notebook specifically for watercolor painting would

A study of lichen and moss on fallen bark (left) *and record of birds during an official count* (right)

be best. And if you don't know what media you're going to work in, a mixed-media book with paper that can handle multiple mediums such as watercolor, pencil, and ink may be most suitable.

Some journals are bound like books, while others are spiral-bound and easy to lay open. Some have perforated pages that can be torn out. Some journals are more flexible and have a basic cover with inserts that can be changed out. Journals come in many shapes and sizes, and choosing one is all about personal preference.

Some naturalists prefer larger books the size of school notebook paper, while others prefer smaller books they can easily carry. Perhaps you prefer to use loose pages to display, share, or organize in your own way later.

If you find that you struggle to actively use your journal, make the journal itself more fun. Add stickers or washi tape to decorate it, or buy a journal with a cover design that makes you want to use it and carry it around.

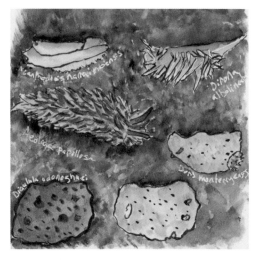

A record of nudibranchs discovered during low tide

Another option is to keep a digital journal on your phone. There are many journaling apps where you can write observations, add photos, record sounds (perhaps even an original song), and even create videos.

Pencils and Pens: Like the journal, these tools are also a matter of preference. A pencil and eraser are essential for quick sketches and layouts, and you may want to keep your journal minimal and use only pencil. The types and styles of pens are even more numerous than journals. If you're new to art or journaling, choosing one can be intimidating. Consider getting a set of pens of various thicknesses in the beginning. Tips are usually measured in millimeters and come in standard sizes, from super fine to thick. Pay attention to whether the ink is water based or not. If you want to mix pen and watercolor, don't get a water-based pen because it will run and bleed when you paint over it. Look for permanent-ink pens.

In addition to basic drawing pens, there are many brush pens options. Instead of a solid or rollerball tip, these pens have brushes, which offer the ability to draw in various thicknesses. The flexibility and adaptability of brush pens reflect the forms of nature and can work wonderfully for those subjects.

Coloring Tools: Some naturalists color their journals with colored pencils or watercolors. There are many brands of colored pencils, and artists have strong feelings about their favorites. If you're a new artist, don't spend a lot on these products

Documenting moss life observed under a microscope

right away. Consider buying student-quality tools instead of professional-quality ones at first, to experiment and see what medium works for you. Try purchasing a few single colored pencils instead of an entire set to test out the brand.

You can find many small, travel-sized watercolor sets that easily fit in field bags. A handy tool for creating watercolors in the field is a waterbrush, which is a paintbrush where the handle stores water and feeds it directly to the brush so you don't need to carry an extra container for water.

In addition to standard pencils or watercolors, there are many options that merge the two. Watercolor pencils look like regular colored pencils, but when you paint over them with water, the color spreads and acts more like watercolors on the paper. Similarly, watercolor brush pens look like markers, but once you add water to the paper, the ink acts like watercolor. Both can be used on their own or in combination with other mediums. The illustrations in this book were all done with an ink brush pen, watercolors, and watercolor brush pens.

Other options include charcoal and pastels, but these have to be sprayed to prevent them from smearing. If you want to include bark or leaf rubbings in your journal, carry pastels or a crayon. A white pen or white brush pen can be useful to add highlights.

Ruler: A small ruler is useful not only with your art supplies but in your field bag too. It will help you measure items you encounter, take photos for size documentation, and draw perspectives when you want to create straight lines. Get one with metric measurements because most scientific references use the metric system.

Wildlife seen on a visit to a state park (left); blooming plants at a city park (right)

Adhesive Spray or Tape: If you want to put pressed plants directly into your journal, you can use adhesive spray at home, or tape, which is more portable in the field. It's also possible to glue or tape drawings done on loose paper into your nature journal.

Steps

Choose a Location: You can journal about nature almost anywhere—in your backyard, on a rooftop deck, in the forest, or at the seashore. You don't even have to go outside; you can sit at home near a window and document what you see, or journal about plants, terrariums, or other indoor projects found in this book.

Add Relevant Information: Before you start, make a note of any relevant information you want to include for this journal page and for the nature journal as a whole. You could include the date, time, location, and habitat type; details about the weather, like cloud type, temperature, or wind direction; or things like the phase of the moon and tide height. You can use a consistent format for all the pages in your journal or take a more free-form approach. It's entirely up to you, but before you dive in, consider the journal as a whole.

Observe: First, forget about your journal and simply look around. Take time to settle in and quietly observe using all of your senses. Notice as much as you can, but also close your eyes and try to pay attention to each sound you hear. Take deep breaths to discover what scents you can detect and where they're coming from. Become aware of your sense of touch, and feel the sun or mist on your face or arms. Experiment with your sixth sense as well. Do you feel a sense of danger? Peace? Unease? Wildlife is very in tune with this sense, and with practice and exposure, you can pick up on it as well (see Project 19: Learn the Language of Nature).

Try to form a complete picture of the habitat you are in, especially if you're taking a photo to draw from later at home so you

> ### Tip
>
> When starting a new journal, skip the first page or two and leave them blank. This approach can relieve the pressure of making a perfect first page. Plus, once your journal is done, you may wish to add a table of contents or some other overview of its contents on those first pages.

can remember the sense of place when you've left. Once you get a feeling for the location as a whole, focus in on smaller pieces. You don't have to notice everything, just whatever catches your attention. Maybe it's the way a hummingbird is preening on a branch, or the tracks of a leaf miner on a single leaf. Perhaps you've found a slime mold plasmodium growing on a log.

Write: Empty pages can be intimidating. To break in the page, add a few basic notes about the location, weather, time of day, and any other relevant data. You may choose to make a list of everything you see, hear, or smell or record your first impressions of the place. If you observe specific animal behavior, describe in detail what you witness. If you see an unfamiliar plant or animal, write down all the information you can, like color, texture, and size, to help identify it later. It's always better to have too much information than not enough. Write down questions that come to mind while you're observing. You can also try to write a stream of consciousness or express yourself with a haiku.

Draw: Consider what you want to capture on that specific page. Maybe you want to include a drawing of the landscape to document the habitat, or perhaps you want to focus solely on parts of a specific tree.

Draw it in the field or take a photo to draw back at home. If you take pictures, shoot several from different angles to capture the subject as thoroughly as possible, and make notes in your journal or phone so you won't miss important features later. Photos often don't capture color or tiny details well.

If you're at home drawing items you collected, study the objects not only with your eyes but also with your other senses. Pick it up, feel it, smell it, and listen to the sounds it makes. Maybe it rattles or makes a hollow sound when tapped with your finger. Add these observations to your journal as well.

Start with a quick pencil sketch to establish the rough layout, leaving room for your notes and other writing. If you're using pen, trace your pencil lines in ink and wait for it to dry. Then erase the pencil and add color, if you're using it. If you're not using ink, you can dive directly into your watercolor or colored pencils. Pencil lines often show up through watercolors, and many artists prefer that, but if you don't want your sketch lines to be visible, you can sketch the scene first in watercolor pencil instead of a graphite pencil so it will blend into the painting.

PROJECT 2

Go Pond Dipping

SEASONS:
All

STUDY TOPICS:
Freshwater invertebrates,
pond life

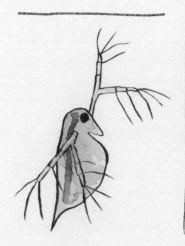

When you gaze at a pond, you may notice a motionless heron hunting in the cattails or a small fish lazing in the shallows, but there is also an entire, nearly invisible world living beneath the surface. You don't have to put on scuba gear to find out what's going on down there though. There's a much easier, fun way to explore the life of a pond.

Pond dipping uses simple tools to temporarily scoop out invertebrates, or animals lacking a backbone, and other living creatures to observe and document. You can pond dip year-round, and while spring and summer will produce the most wildlife, in the cold months, invertebrates like dragonflies live down in the depths of ponds covered by thin layers of ice that you can break through to investigate.

Water striders near a water lily (above); *water flea* (previous)

No two ponds are the same, and they differ dramatically around the world. You may decide to explore a mountain pond formed by a glacier, a marsh with bog moss covering the edges, a pond left behind by a stream that changed course, or an artificial, concrete urban pond. Each pond, large or small, is a complex habitat determined by light and water temperature, which varies depending on the season, oxygen levels, and mineral content, among other factors. These factors play a role in the layering of the pond, what plants grow within the pond, and what animals can be found in it. Every pond will change throughout the seasons, often dramatically. Temperate ponds may develop a layer of ice in winter, while in summer, water flea populations may explode in some ponds.

A pond may look simple from the surface, but underneath is a complexly layered habitat that varies. Algae and invertebrates such as water striders or whirligig beetles can be found at or right under the surface, called the "surface film" layer. Drifting animals such as zooplankton will be in the open water where rooted aquatic plants do not grow. Around the edges is the littoral zone. This area is where invertebrates that require shelter, like larvae of dragonflies and damselflies, diving beetles, and mayflies, will be found amid the rooted aquatic plants. At the bottom of the pond, burrowing invertebrates such as crayfish, snails, bloodworms (fly larvae), and different dragonfly larvae live where plants do not grow.

All the activity in a pond is connected. If more algae grows, the water flea population may respond accordingly. If the water flea population grows, then the invertebrates that eat them will likewise flourish, attracting more fish and birds.

The time of day affects many ponds as well, and in some ponds, there is a complex daily vertical migration up and down the water column. Zooplankton spend the night near the water's surface, but at dawn they begin to move down into the depths where they spend the day. Ponds also vary from year to year, and they change as they age. New ponds have a clear bottom, but the composition changes dramatically as plants grow and die, laying down layers of decaying matter. Eventually, as the aquatic littoral plants keep growing in toward the center, the shoreline shrinks. The aquatic plants die and decompose in the bottom, and slowly, the pond gets shallower and shallower until it becomes a swamp. As the plant matter completely fills in, the pond eventually turns into a meadow.

Materials

Net: The only thing you might not already have at home for a successful pond-dipping expedition is a net, but you can find inexpensive ones at pet supply or aquarium stores or online. A simple aquarium net is small, easy to carry, and works very well for dipping around the edges of a pond. To dip deeper or farther out, look for a net with a longer handle.

You'll want a net with densely woven mesh so small invertebrates don't slip through the holes. To catch zooplankton, look for a special net with a dense weave of 180 meshes per inch, which you can find at online entomological supply stores.

Container: You'll need something to put the invertebrates in where they can safely rest or move around while you observe, make notes, or take photos. Any shallow container with a flat bottom, like a pan, bowl, or plastic food container, can work. Look for used containers at thrift stores and give them a second life. Ice cube trays are also useful for

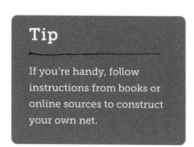

Tip

If you're handy, follow instructions from books or online sources to construct your own net.

Tip

Try not to scoop up the substrate of the pond's bottom with your net. The mud or rocks won't filter through the mesh, and the water won't drain well.

putting individual invertebrates in for observation or organizing. It's easiest to see the invertebrates if the container is white.

You can also shine a light under a large, clear petri dish—simulating how a microscope works—to view the insides of some invertebrates such as water fleas or mayfly larvae.

Hand Lens: Once you find invertebrates, a hand loupe or magnifying glass can help you view them much better and reveal details like the gills of mayfly larvae or the eyes of a diving beetle larva.

Pipette: A common laboratory tool, a pipette is like a small turkey baster and excellent for moving tiny aquatic organisms. There are many inexpensive options of pipettes that will work for naturalist activities. Even the ones labeled "disposable" can be reused several times.

Microscope (optional): To view microscopic life you will need a compound microscope. If you want to get an intimate look at the larger aquatic life like invertebrates, algae, or plants, you'll want a dissecting microscope.

Steps

Find a Pond: You don't need to look for a pristine mountain pond to find invertebrates. Any body of water you encounter—from a backyard pond, rainwater detention pond, or concrete fountain to a mountain pond or wetland—will have a unique ecosystem. Even seasonally temporary ponds, called vernal pools, host a unique collection of invertebrates. Be sure to explore different ponds, since each one has a different ecosystem.

Fill Your Container: Scoop up some of the pond water and pour it into your viewing container. Set it aside. Make sure it's shaded from sunlight if possible so the shallow water doesn't heat up too much and harm or kill the invertebrates.

Dip Your Net: Simply sweep your net along the edges of the pond. If there is aquatic vegetation, move your net through it to find damselfly larvae and other invertebrates. Open water can yield water fleas and diving beetles. Surface algae is also a great place to look. Scoop out a handful and pick through it methodically to find all manner of invertebrates.

Empty Your Net: Hold the net over your container and turn it inside out. Dip it into the container water to wash off any invertebrates clinging to the net. If you want to separate individual invertebrates, use a spoon to scoop them into the compartments of an ice cube tray. For smaller organisms, a pipette will work well.

View: It's time to observe! Many organisms will quickly become obvious as they move to try and find shelter. Others will lurk, and you may have to sift around to encounter them. Watch the different, creative methods the invertebrates use to move. Observe their anatomy and note whether they use gills or another way to breathe underwater. Study their eyes, mouths, legs, and antennae.

If you live near a pond, you can take a jar of pond water home to view under your microscope. Many smaller organisms, like diatoms and ciliates, are found in virtually every body of water but are only visible with a microscope. Diatoms are single-celled algae that come in a variety of shapes, with rods being a very common form. Their cell walls are made of opaline silica, so they are basically algae living in a glass house. Ciliates are protozoans and also have a wide diversity of forms. Their most notable features are their hairlike appendages, called cilia, which help them move, feed, and even sense things. Be sure to return the water to the original pond when you're done so that these small organisms can resume their lives. Introducing potentially new organisms to a different pond can disrupt the aquatic ecosystem. Or you can use it to inoculate your wetland in a jar (Project 3).

Identify: Invertebrates are incredibly challenging to identify down to the species level, but with exposure and practice, you can begin to get a sense for different groups of aquatic organisms and learn to categorize them to at least the family level, maybe even genus. You can take a pond life field guide (see the "Pond Dipping and Wetland in a Jar" section in Resources for recommendations) with you, or take photos or make sketches of the organisms you encounter in the field to compare to the ones in your

Tip

You can take macro, or very close-up, photos with your phone either by using a special attachment that clips on the phone over the camera lens or by simply holding a hand loupe over the camera lens.

books when you return home. It's also beneficial to study field guides before pond dipping to get a sense of what the various invertebrates look like.

Document: Write your observations in your nature journal, sketch the invertebrates, or take photos or videos of what you've found.

Pond Life

You will probably encounter quite a variety of aquatic life during this pond-dipping activity, depending on where you live. From single-celled animals, known as the protozoa, to rotifers and hydras, diving beetles, stonefly and caddisfly larvae, worms, snails, scuds, crawfish, and spiders, you never know what you might find. Aquatic invertebrates are incredibly diverse and encompass multiple groups, including insects, worms and other arthropods, mollusks such as snails, and arachnids such as mites and spiders. Here are some common organisms found in ponds around most of the world. A few more are highlighted in the next project, Create a Wetland in a Jar.

DRAGONFLY LARVA (ODONATA)

One of the fiercest predators in a pond, the dragonfly larva will eat anything it can catch. Some species lie in wait for something to happen by, and others actively hunt, but all are deadly. They have an explosive lower lip called a labium that is held shut with a locking mechanism. When they find prey, they release the mechanism, and their jaw shoots outward at incredible speed while palpi, which act like meat hooks, grab the prey and pull it back into the dragon's mouth.

Dragonfly larvae are strange-looking creatures, usually brown with six legs, hairy in some genera, and spindly in others. A few genera are long-bodied, but many are quite stout and sometimes flat depending on their lifestyle of actively hunting or lying prostrate. Some dragonfly larvae swim, but most crawl along through plants and algae around the pond's edges or lurk on the bottom, sometimes buried in mud waiting for prey. Dragonfly and damselfly larvae are similar, but dragons tend to be rounder and more squat, while damsels are long and slender and have three gills at the end of their abdomen that resemble feathers.

WATER MITE (HYDRACHNIDIA)

One of the only arachnids—the family that includes spiders, ticks, and mites—found entirely in fresh water, water mites swim beneath the surface of ponds. They generally have a plump, round body with eight segmented legs and pointed mouthparts on the face with pedipalps, their small, drumstick-like sensory organs. They're tiny, only 2 or 3 millimeters in size, and can be surprisingly colorful. Some are bright red or orange, while others are brown, but you'll recognize them as water mites because adults have eight legs, like all arachnids.

There are many species, but most can be found in similar habitat and live near the bottom of shallow ponds. Mites are also parasitic as early stage larvae and sometimes as adults, attaching themselves to other aquatic invertebrates to feed off. They are so abundant in water bodies that mites typically parasitize 20 to 50 percent of the invertebrate hosts in a particular pond. Most can swim, and some crawl on plants or along the substrate. Nearly all are predators, hunting fly larvae, mosquito larvae, and other small aquatic invertebrates by grabbing them with their pedipalps. The mites pierce the prey they catch with their sharp mouthparts and inject enzymes produced in their saliva, which break down the insides of the prey. After the enzymes dissolve the prey's body, the mite sucks out its insides like a milkshake.

Water mites have a life history so incredibly complex that science still doesn't have a determined set of terms for the many stages from egg to adult.

BACKSWIMMER (NOTONECTIDAE)

Backswimmers are the upside-down version of water boatmen. Both are similar in appearance: they are oval-shaped and usually brown or dark red with one pair of long legs that spread out like boat oars. The second pair of legs is not nearly as long but longer than the third pair, which they fold up by their head, ready to grab prey. If you look closely, you'll notice their two large red eyes.

As their name implies, they swim just under the surface of fresh water on their backs using their long, hairy hind legs to row themselves through the water. As they swim, their head tilts down, and the tip of their abdomen touches the surface. There's a reason for that position. Being insects, they need to breathe air, but these backswimmers have a special method of trapping air in hairy troughs on the sides of their abdomen, as well as under their wings when they're at the surface, so they can breathe while diving in search of prey.

Backswimmers are predators and will consume anything they can catch, usually insects and crustaceans, but also small tadpoles and even little fish. They have piercing mouthparts they use to stab their prey and suck out their body fluids while they hold on to it with their spiny front legs. Although they may not look like it, they are strong fliers and will disperse to search out new ponds to colonize.

MIDGE LARVA (CHIRONOMIDAE)

The midges of this family are of the nonbiting variety and are an important source of food for a variety of bigger life in the pond—and then as adults, for life out of the pond, like birds. Nonbiting midges are the most diverse group of all aquatic invertebrates and also abundant, with reports of as many as fifty thousand individual mites per square meter. Midge larvae can be found throughout all the microhabitats of a pond. They can account for half the species in a pond, so there are likely some in your pond-dipping container.

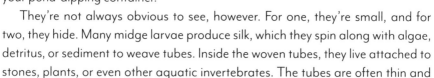

They're not always obvious to see, however. For one, they're small, and for two, they hide. Many midge larvae produce silk, which they spin along with algae, detritus, or sediment to weave tubes. Inside the woven tubes, they live attached to stones, plants, or even other aquatic invertebrates. The tubes are often thin and very fine, and if you look, you can see the midge inside as they wiggle to draw water in. All midges are herbivores and generally consume organic matter. Some also use their silk to capture their food in a net they construct.

At first glance, they resemble worms, but they actually have tiny prolegs (fake legs) at both ends of their bodies. They have segmented bodies but are otherwise nondescript. Midge larvae are common but highly varied and come in many different colors. One particular red midge, known as the bloodworm, is red because of the hemoglobin in their blood, a feature that allows them to survive in ponds with low oxygen levels.

WATER STRIDER (GERRIDAE)

Skating along on the pond's surface, slender, chocolate-brown water striders are specially adapted to use surface tension to walk on water. Their most notable feature is their extraordinary legs. Water striders have four long legs, and two smaller ones that are tucked up under their heads, which they use to gracefully glide over the surface. They're agile on the water, skating across the surface because of special hairs on the ends of their legs that repel water. Their claws, called tarsus, are also found higher up on their legs so they can avoid touching and piercing the water's surface.

Water striders detect prey by sensing vibrations on the water with their legs. When they find prey, they use their smaller, tucked-away legs to grab and hold it raptor-style. Water striders will eat any other invertebrates they encounter at the surface, alive or dead, including aquatic insects that venture near the surface, or hapless insects that fall into the water. They will even prey on other water striders, since species frequently congregate in large numbers.

Create a Wetland in a Jar

SEASONS:
All

STUDY TOPICS:
Wetland ecology, freshwater invertebrates

Marsh, swamp, bog, fen, mire, muskeg—wetlands have many names. That's largely because this type of landscape is not uniform; there are many different types around the world. Bogs are unique for their sphagnum moss, a genus of bog moss that eventually decomposes into peat. Swamps have woody vegetation, like alder and cypress, while marshes have soft-stemmed vegetation, like cattails and reeds. Many wetlands have freshwater and saltwater variations as well. Each type of wetland serves as vital habitat for communities of organisms, comparable to rainforests as productive habitats, and is tremendously important as a source of food for a wide range of wildlife.

Wetlands are dynamic landscapes that change dramatically with the seasons. Some

A wetland in a jar

may dry up completely in the peak of summer. Others are a welcome source of water for wildlife throughout the year. But all wetlands host a wealth of diversity, serving as habitat to mammals, birds, reptiles, and amphibians, as well as a large range of invertebrates. They provide temporary stops during migrations, a source of food in winter, or a nursery to raise young. For many species, a wetland is a permanent home.

Humans also depend on wetlands because they are nature's sponges. In the face of climate change, they're as important as ever, storing a large amount of carbon. Yet over the last three hundred years, the world has lost 87 percent of its wetlands. Without them soaking up and holding the deluge of rainwater, our cities and towns could also flood more quickly. At the same time, wetlands hold rainwater and release it slowly, serving as a source of water during dry periods and drought. A mere acre of wetlands can retain an astonishing one million gallons of rainwater. Wetlands also act as water purifiers, like the kidneys in your body, filtering out pollutants before they end up in streams and rivers. Without wetlands, our creeks and rivers would quickly fill up with sediment that runs off the landscape during rain. Wetlands intercept the rain and keep our waterways running smoothly.

Wetlands are a miracle landscape, able to help mitigate the effects of climate change and reduce flooding, pollution, and drought. Plus, they provide essential habitat. But not only that, they also offer humans a place for recreation, to enjoy nature walks, kayak or canoe trips, and bird-watching. For naturalists, wetlands provide the opportunity to observe and study this diverse landscape.

Red-winged blackbirds trilling on the cattails are easy to see. The chorus frogs singing in the spring are easy to hear. Dragonflies buzzing over the pond can't be missed. But how often have you watched a dragonfly larva? It's harder to observe the diversity of life beneath the surface of the water in a wetland, or even know what lives there, but it's not impossible. There is a way to not only see what lives under the water but also bring it home for long-term observation and study in a miniature, re-created wetland.

Once established, a miniature wetland will be self-sustaining, giving you the chance to observe the life cycle of different invertebrates over a period of time. You can watch snails mate and lay eggs, and then observe the tiny, glass-like baby snails emerge. You can see water fleas swim with their disproportionately long, fingerlike appendages and the hydra clone itself. You can even use your wetland jar as a temporary observation aquarium for larger invertebrates that can't live their entire lives in a small container, like dragonfly larvae or diving beetles. If you find a particularly interesting invertebrate while pond dipping, you may want to bring it home, place it in the wetland jar, and observe it for a few hours or a few days before returning it to its original wetland to continue living.

And the best part? Unlike setting up a fish aquarium, creating a wetland container is very simple and requires few materials. You won't need to buy pumps or other special equipment—you probably have many of the materials on hand.

Materials

Jar: The first thing you'll need is a container for your miniature wetland. It can be a large jar or any type of glass container with a wide opening to get materials in and out and also for maintenance. It's also better to have a larger water surface area for air and gas exchange to let the water and life in it breathe. Look for round containers at thrift stores at least the size of a large mason jar. They can be taller than wide or wider than tall.

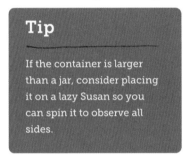

Tip

If the container is larger than a jar, consider placing it on a lazy Susan so you can spin it to observe all sides.

Soil: Regular garden soil works best for this project. Don't use fancy potting soil, which has too many nutrients in it. If your only option is potting soil, look for organic soil and avoid perlite and fertilizers.

Gravel: Use clean gravel with a diameter of 2 to 5 millimeters. Gravel that's too large or too small won't allow the soil to breathe and can kill the roots of your plants. Sand isn't ideal, but it can work.

Water: You'll need enough water to get your wetland started and also for keeping it full once established. Tap or rainwater both work, but rainwater is easier on the ecosystem.

Aquatic Plants: Wetland plants that grow entirely underwater or are emergent (growing in water but breaking the surface) will both work for this project. You can collect these from ponds (with permission from the owners if the pond is private). Many aquatic plants can be propagated with just a small piece of the plant, even without the roots. Look for plant pieces that wash up on the shore of local lakes, but consult a book on wetland plants of your specific area to make sure they aren't invasive. You can also visit aquarium shops to buy aquatic freshwater plants, although you're less likely to find native species.

Tweezers or Pliers: To place the plants in the soil, you'll need a tool like tweezers or pliers.

Rubber Tube (optional): If you choose to siphon the water into the container, you'll need rubber tubing.

Hand Lens or Magnifying Glass (optional): If you want a closer view of the life in your wetland, a hand lens or magnifying glass can help you see them easier.

Compound Microscope and Slides (optional): If you want to study the microscopic life in your pond water, you'll need a compound microscope and concave slides with cover slips.

Pipette (optional): To transfer pond water to a slide, you'll need a pipette.

Steps

Layer: First, put 1 to 1.5 inches of soil in the bottom of the glass container. Over the soil, lay an equal layer of the gravel 1 to 1.5 inches deep. If you choose to use sand, it needs to be a thinner layer than gravel, only 0.5 to 0.75 inch deep, so the soil can get air.

Add Plants: Use tweezers or pliers to push the plant roots, or stems if you have cuttings, into the soil. Don't be afraid to experiment with plants. If they die, try something else, and if they do too well, remove them or cut them back.

Place the Container: Where you place your container is important. Aim for a consistent temperature and good, indirect light. Windowsills generally offer reliable sunlight, but the temperature fluctuates too much during the day, particularly in summer and winter, and can cause your ecosystem to crash. If you can't find a place near a window with natural light, you can place a lamp over the container to shine artificial light on the plants for at least twelve hours a day.

Add Water: This step needs to be done slowly and with care. If you dump the water in, it will disturb the soil and you'll have a muddy mess. Let the water slowly flow in. One method is to use a rubber tube and

> ### Tip
>
> Expect to wait about six weeks until the soil stabilizes before you do anything else. You may see fluctuations in algae and murky water until it settles. The plants will help purify the water as they grow and control algae. You can try inoculating your wetland sooner, but if you do, the creatures you collect may not survive.

siphon in the water. A simpler method is to hold your hand over the gravel and pour the water slowly on your hand to break the flow.

> **Tip**
>
> If the container becomes overgrown with algae or the plants die, reconsider its location and check the amount of light and temperature. Snails are also important. If you didn't get any with the first inoculation, collect more pond water and add it to your container.

Inoculate: After about six weeks, when the water appears clear and the fluctuations have stopped, visit a wetland to collect water. Take along a small container with a lid, and fill it up with the pond water. Try to collect a few aquatic snails if you see them on vegetation. When you return home, gently pour the container of pond water into your wetland container. If you're lucky, you'll get an interesting variety of aquatic life in your container. Much of it will be tiny and hard to see with the naked eye.

Observe: When your container is finished, start observing the freshwater invertebrates that live in your small wetland. You can see many different invertebrates with your bare eyes, or you can use a simple hand lens or magnifying glass to study the organisms from the other side of the glass without disturbing the ecosystem. To look even more closely at what's living in your container, take a water sample with a pipette, put it on a slide, and look at it under a microscope. You may find diatoms and ciliates.

Maintain: Once your wetland is stabilized and inoculated, all you'll need to do is add water once in a while. Snails and flatworms collected from the pond will eat algae and keep the container clean, and the plants will purify the water. If the plants grow too dense or too large, trim or remove some so light can filter through the entire container.

Wetland Life

Freshwater invertebrate life in wetlands will vary widely from location to location around the world. However, here are a few organisms that are common nearly everywhere. Check the "Pond Dipping and Wetland in a Jar" section in the Resources for recommendations of field guides with more aquatic life.

SNAILS

As a whole, aquatic snails aren't very colorful. Their shells are shades of earth tones and their bodies similarly subdued. But examining them up close reveals endearing and remarkable features. Like their land counterparts, these snails have tentacles, similar in appearance to antennae of insects. But unlike land snails, aquatic snails have only one pair on their heads. Their large foot will be easy to observe as they glide along the surface of the glass on a layer of mucus.

Even more interesting, though, is observing their tongue-like append-age known as a radula. Snails play an important role maintaining algae levels in almost every wetland ecosystem, as well as in your wetland jar. Most aquatic snails feed on algae by scraping it with their radula. Quite unlike our own tongue, it's hard and covered with thousands of razor-sharp, very tiny teeth. The snails push the radula out of their mouth to scrape and grind algae from the surface, which they bring back into their mouths to consume. Some aquatic snails find the algae on plants, while others filter algae out of the water with internal gills.

If you're lucky, snail eggs may even appear on the vegetation, fol-lowed by teeny-tiny baby snails that are almost clear when they first emerge.

FLATWORMS (TURBELLARIA)

Freshwater flatworms are, as the name implies, flat. They're usually shades of brown or gray and have an arrow-shaped head with two cartoon-looking crossed eyes on the top. These are simple eyespots and only detect light, as flatworms can't see shapes or col-ors. Some species are so small they're micro-scopic, but others can be seen with the naked eye. Flatworms cling to vegeta-tion, or the glass container of a wetland jar, and slide around like snails. Unlike snails, flatworms are predators and feed on small invertebrates. They're not picky about whether their food is alive or dead either.

Flatworms are largely transparent, so you can observe their digestive system. You might notice they have only one opening on their body, a mouth that doubles as an anus. They use a structure called a pharynx to feed by sucking the fluids out of their prey. Even more than that, the flatworm's biggest claim to fame is their incredible ability to regenerate. If they are cut in half, they will not die, but instead turn into two flatworms. Even if they are cut into many pieces, they will usually regenerate into entirely new worms.

HYDRA

You may encounter multiple types of hydra in your wetlands. The most common is a green hydra, but there are also white or gray hydra, as well as brown hydra with very long tentacles. Hydra look like a cross between a tiny sea anemone and a jellyfish, which they are actually related to. There is nothing else like hydra in the freshwater environment, and despite being a relatively simple organism, these small animals have some amazing powers.

Just like their jelly relatives, hydra possess short, stinging tentacles they use to harpoon passing water fleas or other zooplankton with toxins, which they then swallow whole. What goes in must come out, and like the flatworm, the hydra has only one opening—its mouth is also its anus. Hydra largely remain rooted in place by plastering their foot to the surface of duckweed or other aquatic plants, but they can move if they are disturbed or food is scarce. Then, they'll unstick their foot and start to somersault through the water, turning tentacle over foot until they find a more preferable location. Hydra reproduce by budding, which is essentially growing a small clone from their body. You might see a hydra with a tiny one on its side, or sometimes two, almost equally large hydra stuck together. Eventually the clone breaks off and goes its own way.

ZOOPLANKTON

Just large enough to see with the naked eye, zooplankton are the most common of freshwater invertebrates. Most are crustaceans and include water fleas, scuds, seed shrimps, and copepods.

Water fleas are strange-looking invertebrates that dart around with long, fingerlike appendages sprouting out of their round, transparent bodies like wings. They make the freshwater world go round, because water fleas are the base of the food web. Nearly everything starts with them. Their population fluctuates wildly because females can produce young every three to four days, from one to one hundred at a time. No males are even required—females can reproduce by a process called parthenogenesis. When a new group of water fleas emerges, their predators go on a feeding bonanza, and the water flea population drops again.

Scuds are related to isopods and crayfish and have the typical crustacean armor. But they're tiny, no more than 20 millimeters long, and usually white or brownish. They have a mass of legs under their comma-shaped body, which is not round like a roly-poly but flat. Seed shrimps look, as the name implies, like tiny, swimming seeds, with little, feathery appendages poking out. Copepods look like elongated roly-polys with long antennae. They can often be seen carrying sacs of eggs.

Make a Bathyscope

SEASONS:
All

STUDY TOPICS:
Aquatic life in a natural habitat, aquatic wildlife behavior

Curiosity about "what lies beneath" has driven humankind to all manner of extreme solutions to discover what goes on underwater. From giant metal helmets attached to hoses in the earliest days of scuba diving to modern high-tech, deep-sea submersibles, our solutions have always been inventive and complex. But often the simplest tool can act as a magical key for naturalists in opening the door to other worlds.

Three-quarters of our planet is covered in water, whether it's a tide pool or the deepest sea, a mountain river or an estuarine wetland. Lakes and ponds vary widely around the world, with distinct sets of plants and animals that are different from rivers. A pond in Japan differs significantly from a pond in Iceland. But did you know that

A sunflower sea star, clown dorid nudibranch, and little red dorid viewed through a bathyscope

aquatic life can vary dramatically from one body of water to another, even if it's just a half mile away?

A naturalist observing a river may see salamanders and fish, caddisflies and crawfish. Searching through a nearby lake might reveal freshwater jellyfish, dragonfly larvae, or the nest of a stickleback fish. Saltwater offers a massively diverse range of habitats, from tide pools to mudflats, coral reefs to fjords, many of which offer a unique environment where fresh water and salt meet.

Wildlife can be observed with a bathyscope, but let's not overlook the chance to examine plant matter living in its natural habitat, because aquatic vegetation is not easy to observe with a net. Like trees, aquatic plants serve as a source of food, as well as shelter, for wildlife. If you watch closely when it's sunny, you may see aquatic plants release tiny bubbles as they photosynthesize.

What lies under the surface in a body of water is a largely unknown and unseen world that changes throughout the year. Even in the ocean, aquatic life is in tune with the seasonal changes, and all aquatic life reacts to the effects of climate change, pollution, and other human-caused impacts. One particular sea slug, the bright pink Hopkins' rose nudibranch, expands its range north up the Pacific Coast from California into Oregon in years with warmer-than-usual sea temperatures. The nudibranch's movement is an indication and reminder of the effects of climate change on the coastal marine ecosystem.

We've seen how a naturalist can use a net to discover what invertebrates live in the water (Project 2) and re-create a wetland in a container for study (Project 3), but the easiest way to truly observe beneath-the-surface life in its natural habitat doesn't require submarines or scuba gear. Instead, studying life underwater can be done with a simple, basic device accessible to most anyone.

The history of the bathyscope, also known as an aquascope or water telescope, is unclear, but the first one was likely invented shortly after the introduction of glass. It's simply a hollow tube or box of varying length with glass on one end and an open end on the other. Putting the glass end into the water and looking through the other end is an ingenious, incredibly easy way to reveal not just one hidden world but any underwater realm, from fresh water to salt and everything in between.

The bathyscope cuts through the ripples on the water's surface, which can be extreme in environments like the ocean and rivers. But even in calm lakes and ponds, breaking the water's surface can make a stunning difference because it reduces glare, allowing a clear view into the depths. We know from early reports of water telescopes in the 1880s that viewers can see up to 20 fathoms into the depths. That's about 120 feet, or 36 meters—an astonishing distance with an extremely low-tech device.

The water telescope was used largely by fishermen in search of fish shoals, but many people used early water telescopes or bathyscopes to survey a ship's hull for damage, search for sunken wrecks, and even find dead bodies. Naturalists made use of the bathyscope to survey water bodies and observe aquatic life. Their early versions were constructed out of wood, followed by tin and other metals, or even a simple bucket with the bottom knocked out. Most modern versions are made from PVC pipes. Longer tubes, some multiple feet in length, are weighted to counter the buoyancy of the air, but have retained the basic form. Another common form is a funnel shape with the wide end sealed in glass.

You can purchase commercial versions of the bathyscope, but it's easy to construct your own out of repurposed or simple materials. Naturalists who are handy with basic tools can construct a more durable bathyscope.

Materials

Container: There is a huge variety of options for making a bathyscope. In the simplest terms, the underwater viewer needs to have a clear end of glass or plastic attached to a container with an open end opposite the clear end. The container can be tubular or square, linear or conical. You can repurpose large metal cans, plastic takeout containers, ice cream tubs, clear ice buckets, plastic pantry organizers, and even larger storage containers. Basically, anything waterproof and big enough to look through, with either a clear end or one you can cut out, can work—even better if it has handles.

The depth of the container will reflect its intended use. For shallow aquatic areas without heavy swells, like tide pools or ponds, a shallow container at least three inches deep is sufficient. For places with waves or larger swell like oceans, rivers, or lakes, it will need to be several inches tall to avoid getting swamped.

Tip

If you use a plastic container, choose a transparent one without a pattern on the end that would obscure the view.

If you want to construct your own high-quality, long-lasting bathyscope, you can use a wide-diameter PVC pipe, and if you want to add handles, you can use basic drawer handles or rope. To make a bathyscope with a periscope feature, you'll need a PVC elbow joint and a circular mirror the same diameter as your pipe.

Glass or Plexiglass: One end of the bathyscope needs to have a transparent cover, and there are several options for adding one. The clearest option, glass can be found in old frames at a thrift store or purchased and cut to size. Plexiglass is a less fragile option and also easier to cut. For repurposed containers like tins and tubs, plastic wrap can work as a temporary option.

You can glue your material onto the end without cutting it to size. However, be careful with any sharp corners that stick out.

Sealant: If you're constructing your own transparent end, you'll need a sealant to glue the glass to the container and make it waterproof. Look for aquarium sealant because it's nontoxic and won't harm the life in the water.

Black Paint or Tape: For any container deeper than a couple inches and not solid like wood or tin, it's beneficial to block out the light by wrapping the container in dark duct tape or painting it black with nontoxic paint.

Steps

Construct: To make a simple bathyscope with repurposed items like an ice cream tub or takeout container, first cut out the bottom of the container with scissors or a box cutter. Take the lid and cut the inner part away while preserving the rim that clasps onto the container. Stretch plastic wrap across the inside of the lid and attach it back onto the container. If there's no lid, wrap a rubber band tightly around the plastic wrap to attach it. Paint or tape the container if light penetrates it.

Likewise, if you're using an existing container with a clear bottom, like an ice bucket, paint or tape the sides of the container, leaving the bottom clear.

To make a custom bathyscope, cut a piece of PVC pipe or other material to the desired length. If you want to use it on a boat or a dock, your pipe could be up to three feet long. If you want to use it while standing in lakes or other marine environments, you only need one or two feet. It can be even shorter if you plan to stand in deeper water. Too long and it'll be too hard to see into. Paint the tube black if it's not already a dark color. Use nontoxic aquarium sealant to glue the glass or plexiglass onto the end of the pipe. Consider gluing on or otherwise attaching handles, which you can construct from rope or other items found in a hardware store. This will help you hold the bathyscope and prevent it from slipping when your hands are wet. If you're making a long, telescopic bathyscope, you can attach weights to the end to make it easier to handle and counter the buoyancy of the air in the tube.

To make a periscope-style bathyscope, add an elbow joint at the end that gets submerged and glue a circular mirror at a 45-degree angle inside the elbow before adding glass to the end.

Consider making a few bathyscopes of different lengths to use for a variety of applications.

Use: Bathyscopes can be used from floating docks, boats, and kayaks. They can also be used when standing in water like lakes and rivers or tide pools and beaches. Simply place the covered end into the water so that it breaks the surface, and look through the opposite end. You'll immediately enter a new world. Depending on your location, you may find an underwater garden of plants or algae, fish swimming around, aquatic invertebrates crawling over rocks or on plants, or a host of creatures in a tide pool. If you're standing in the water, you'll probably need to stay still until wildlife like fish think your legs are part of the landscape and return to the area after that initial disturbance.

Marine Life

A naturalist could spend countless lifetimes with a bathyscope and still encounter only a fraction of all the aquatic life there is to see. You'll find a variety of marine life you may encounter during saltwater explorations described below. (We covered a lot of freshwater aquatic life in Projects 2 and 3.) Don't overlook the fascinating seaweeds, which are not a plant at all, but algae. The sheer diversity of form and size of the brown, red, and green types of seaweed found around the world boggles the mind. Dead man's fingers are oblong green sacs that grow on intertidal rocks, sticking up like the hand of a corpse, while a kelp can grow into a proper forest, rising up to 175 feet from the seafloor.

Warning

Always be aware of your surroundings while looking through the bathyscope. Look up tide charts beforehand, and keep an eye on waves, boats, and other possibly dangerous factors in the environment.

NUDIBRANCHS

Nudibranch (the name rhymes with *bank*) means "naked gills." Nudibranchs are a type of sea slug, but they little resemble the slugs found in your backyard or a forest. They

range dramatically in size from nearly microscopic to a foot long. When they're out of the water during low tide, nudibranchs look like featureless blobs, but if you observe them under the water, where they are free from the effects of gravity, they transform into magnificent creatures that never fail to wow. Many are flamboyantly colored with spikes, elegant fingerlike projections or bumps that make them look furry. Dorid nudibranchs have feathery or lacy gills, called branchial plumes, on their backsides, utterly unlike that of any fish. Aeolid nudibranchs, illustrated here, carry cerata, which are structures along their backs and can be long and slender or more club-like.

Nudibranchs display a wide combination of colors and patterns. Some are bright yellow; others are pink, red, blue, and even green. A few are white like frosted glass, and others are so translucent you can see their organs. All nudibranchs have rhinophores that resemble squishy horns or antennae on their heads. Rhinophores are as diverse as the nudibranchs that carry them—some resemble lace, others satellite dishes. They are sensory organs, but what exactly they detect is still a bit of a mystery.

All nudibranchs are predators and feed on a variety of other marine life from sea anemones to sponges and hydroids. Most have radula, the scraping teeth common of slugs and snails, but a few in the *Melibe* genus have hoods that filter out food to consume it whole. A number of Aeolid species can ingest a sea anemone's stinging cells, called nematocysts, and then pass them through their body up into the cerata, where the nudibranch uses them to sting any potential predators themselves.

BARNACLES

Barnacles are a reliable organism at many beaches around the world, but are largely overlooked because they tend to blend in with the rocks and become part of the landscape. At low tide most beach explorers don't even notice barnacles, perhaps with the exception of goose barnacles, but observed underwater, they are fascinating to watch.

After spending their larval days drifting in the currents, a barnacle eventually settles down, never to move again. They cement their head onto a rock or other surface, build a protective test—a volcano-shaped shell—around themselves, and spend the rest of their days standing on their head. When the tide is in and they are underwater, they open their plates, stick out their legs, and filter edible particles from the water before retracting their legs with the food attached. Then they do it again and again.

A barnacle's legs are made up of a number of long, slender appendages called cirri, which look feathery. Some barnacles spread out their cirri like a fan and swivel back and forth in the tidal current, catching organic particles like a net would. Other barnacles live as parasites in crabs by infiltrating the crab's armor and taking over its body. The barnacles castrate the crab and then use its body to tend to the barnacle's brood.

The barnacle's biggest claim to fame, however, is that they have the longest penis relative to body size in the entire animal kingdom—up to eight times their body length. Given that they are stationary creatures that can't go looking for a

Hydroid, brozoan, and sponge

mate, but instead have to try and reach a neighbor, this anatomical advantage makes sense. Keep an eye on those barnacles as they feed because there may be more than legs waving around in the water.

HYDROIDS, BRYOZOANS, AND SPONGES

These three organisms are all colonial, meaning what we see as an individual living being is actually a collection of tiny animal organisms. They are unrelated to one another, but even marine naturalists can easily confuse them. They are all filter feeders, meaning they filter and consume phytoplankton out of the water, and they frequently grow in the same habitats as one another, often settling in to live their sedentary lives side by side.

Hydroids, a favorite food of many nudibranchs, are in the cnidarian phylum and related to jellyfish and sea anemones. Nearly all are marine dwellers, with a couple of exceptions, one of them being the hydra we met in Project 3, when we created a wetland in a jar. They often superficially resemble seaweed, with a plant-like form attached to a stable surface such as a rock or dock, but like their relatives, hydroids possess stinging cells. If you look at them closely when they're underwater, you can see the tiny tentacles of each individual polyp.

Bryozoans are also colonial animals and commonly referred to as "moss animals," but they take on many different forms. Some grow upright, while others are encrusting and grow on seaweed or rocks. Their colors range from green and brown to orange and red. Even their textures vary, and some can be gelatinous, while others are very stiff. The individual animals that compose the mega colonial condominium that is a bryozoan are called zooids. Nudibranchs consume bryozoans, and there are a couple of species that spend so much time on their preferred bryozoan prey they have evolved a body that perfectly mimics the color and pattern of their prey.

Sponges also blur the line between plant and animal. They lead stationary lives and grow most commonly on rocks, usually fairly flat, closely hugging the surface, but some grow upward a little. Many are brightly colored with hues of red and purple. Most sponges have easy-to-see holes called oscula. Shaped like small volcanoes, these oscula connect to a canal system that runs throughout the inside of the sponge and filters food and oxygen. Many sponges are the favorite food of numerous nudibranchs, and some species are so close to their sponge prey that not only have they evolved to match the sponge color but their eggs have as well.

Preserve Animal Tracks and Signs

SEASONS:
All

STUDY TOPICS:
Animal signs, mammal surveying, leaves and insects

A naturalist can spend a lifetime exploring an area, and still some animals will elude them. No matter how patient you are, how quiet, how dedicated and observant, certain animals will slip by unseen. Some are shy and avidly avoid humans, others are active only at night, and yet other animals roam such large home ranges that they seldom pass through. But even the "invisible" animals leave signs behind that a naturalist can find. You can find proof of their existence in scat, signs of feeding such as chewed-up acorns or mushrooms, beds and dens, territorial marks like a clawed tree, and tunnels, holes, and other evidence of burrowing.

The most easily found and identified sign of an animal's life in an area is their track.

A variety of mammal tracks, including mountain beaver, meadow vole, northern flying squirrel, junco, red fox, and black bear

These are usually footprints—but not always. An otter, for example, may leave behind a long line from sliding.

Tracking animals' footprints and other signs is an ancient skill and particularly valuable to a naturalist. These ephemeral prints appear when nobody is around, and while the animal itself may elude us, a single footprint encompasses a wealth of information. In fact, scientists have found that some animal footprints are the equivalent of a human fingerprint and can be used to identify an individual animal.

You do not need to drill down to that level of identification with animal tracks in order to learn a lot from the prints left behind. When you take the time to look down, whether it's at old tracks dried in mud or more recent ones, tracks can tell a story. You can learn how fast an animal was moving, how big they were, and how much they weighed. You can follow the tracks of both a rabbit and the coyote that stalked it. These stories connect, overlap, and interweave, telling an elaborate tale about the animals' behavior and interaction with each other and the larger environment.

When you are tracking, often the first step is to learn about the individual footprints of species. If you study types of tracks from a field guide, you can be ready to identify tracks the next time you encounter them.

There are several ways to study animal tracks. You can photograph them, but a photo often cannot capture the depth and fine details of a print like hairs. You can also sketch them and take precise measurements. However, some things are best studied not with eyes alone but also with touch. That's why a common technique for studying animal tracks is to make a cast of them in the field to study at home. These casts capture all the details of the track and are three-dimensional, allowing you to explore with your eyes and your fingers the footprint left behind by a real animal, whether it's from a bear, wolf, or mouse.

Making plaster casts of tracks is a passive approach that can be used anywhere, anytime that you find a track in the wild. A sand or ink trap is a more active method that involves bait and a way for an animal to leave their prints in a place where you choose. For a sand trap, you lay down a simple bed of sand with bait in the middle. This portable approach is reusable and easy to replicate. If you set it up at home, it's also easy to then make a plaster cast from these prints.

The ink trap similarly involves bait, but instead of sand, it uses ink and paper. Animals walk across nontoxic ink, getting it on their feet. When they've finished their meal and walk away, they leave their inky footprints behind on paper. This method is suitable for small mammals, anything up to a skunk or racoon. Mammals are not the only ones who may leave tracks on an ink trap; you may also find signs of passing invertebrates like beetles and slugs. This method is also reusable to an extent and creates a record that is easy to store.

Materials

Plaster of Paris: Basic plaster of Paris is the most common material used to make track casts. Also known as gypsum plaster, this

> ## Tip
>
> Plaster won't work in snow because it gives off heat and melts the track before it hardens. There are specialty products available online made specifically for casting tracks in snow.

powder is used in a wide variety of applications. If you've done any home DIY projects, you may have encountered it before. A milk carton size of plaster costs only a few dollars and will last you for many tracks.

Frame: While it's possible to pour plaster directly onto an animal track, creating a frame around it will allow you to make a thicker and more durable cast. This frame can be made from a short tin can with the bottom cut out, a cut section of a plastic jug, or some other repurposed materials. The simplest frame is a strip of cardboard rolled into a circle like a cookie cutter and held together with a paper clip. It is adjustable for different-size prints, uses common materials, and is easy to carry in a bag.

Ink: To make an ink trap, you'll need a nontoxic material since it will end up on the feet of wildlife. The best option is a carbon powder mixed with vegetable oil, which creates a nondrying ink. You can find basic carbon powder online. Look for carbon or activated charcoal powder that is labeled as all-natural or food grade, and avoid specialty carbon powders that may have ingredients that are harmful to wildlife. There are other ink options, like powder poster paint, but research the ingredients to be absolutely certain they are not toxic. If in doubt, do not use it.

Cardboard: For an ink trap, you'll need a piece of cardboard that you can fold into three sections to make a triangular tunnel large enough for the animal you're targeting. The tunnel needs to be long enough to fit two pieces of white paper at each end with space in the middle for the ink and a bowl of bait. A commercial version is roughly 30 by 50 inches unfolded, but you can customize it for a wide range of animals from mice to raccoons.

Paper: Letter-sized white paper will capture footprints in the ink trap, or cut whatever paper you have on hand to fit your tunnel.

Masking Tape: The wider the tape, the better. You'll need it to tape the paper down and create an ink pad.

Pet or Bird Food: To lure wildlife into the ink trap tunnel or your sand trap, you'll need to research what your target species eats. Or

> ## Tip
>
> If you carry materials in your field bag, seal the dry plaster mix in a secure container so it can't spill and make a mess. Carry water, a plastic container to mix the plaster in, and your preferred method for a frame, and you'll always be prepared to make a cast in the field.

if you just want to see what's around without targeting anything specific, you can try different types of bait. Bird food in either seed or suet form can work for many mammals. Wet pet food typically works well, but if you're using dry pet food, let it soak for a little while first. Other options are peanut butter, nuts, or fruit.

Sand: You'll need fine-grain sand with clay, like bricklayer's sand found at a hardware store, for a sand trap.

Steps: Plaster Cast of Track

Find Tracks: You can go out looking specifically for tracks, but it's also a good idea to carry materials in your field bag for when you encounter tracks on general explorations. Look for prints in soft surfaces like sand and mud. You can wait until the ground is damp or look in places that are usually wet, like river and lake edges, wetlands, and beaches.

Prepare: Clear away any leaves or other debris from the track carefully without disturbing the track. Next, set your frame around the mark left by the animal, sinking it into the ground slightly. Mix the plaster with water and stir it with a clean stick, letting the air bubbles rise before pouring. It should be creamy, not lumpy and not too thin, similar to the consistency of a milkshake.

Make a Cast: Slowly pour the plaster into the footprint, letting it fill up the surrounding frame. It's possible to get a cast with a thin layer of plaster, but the thicker it is, the less likely it is to break. An inch or two is a good depth. Let it set without disturbance for about a half hour until it feels hard and has the sound of ceramic when tapped. Finally, remove the frame, dig up the cast gently, and put it in a bag. Don't try cleaning the cast yet because it has not finished setting.

> **Tip**
>
> Making a cast takes practice. Allow yourself time to get the hang of it. The key lies in the consistency of the plaster mix for individual tracks. Tiny tracks need a thinner consistency so the mix doesn't smash the print. Generally, though, if the mix is too thin, it'll take too long to harden because of the higher water content, and if it's too thick, it will harden too fast and not spread out into the full extent of the track.

Leaf Cast

Not all animal signs come from mammals. Invertebrates leave a great many signs in the wild that you can learn to recognize as well. Leaves are one of the easiest places to look for signs of invertebrates because so many of them consume or otherwise use leaves, such as larvae of moths and butterflies and sawflies. Some moths tunnel into a leaf to consume it. Leafcutter bees cut perfect holes out of the leaf's edge, then chew up the leaf and use it to divide the cells in their nest tunnels.

If you're interested in studying invertebrates, you can capture these damaged leaves as a cast. The leaves could also be pressed, but a cast is more durable for handling either for studying or education. It can also be painted with a glaze to make into a work of art.

FIND LEAVES: If you happen to notice curious markings on a leaf, collect it and keep it safe from damage until you are ready to make the cast. It may help to study a book on insect damage so you can recognize leaves that have been impacted by insects.

PREPARE MATERIALS: Clean the leaf of any attached debris. Prepare a surface by laying down wax paper or plastic. If the leaf is thin or fragile, rub petroleum jelly over it so it's easier to separate from the plaster when it dries.

Mix the plaster powder with water in a plastic container until the mix is the consistency of a milkshake. Let the air bubbles rise.

MAKE THE CAST: Slowly pour the plaster onto the wax paper until it's as large as the leaf. Press the underside of the leaf, which is the more textured side, into the plaster. With your finger, gently press the leaf so it's entirely against the plaster. Any section of the leaf not touching the plaster will result in a blank space.

Let it set for about fifteen minutes, and then peel away the leaf. Let the plaster set for a day to fully harden.

Once it's hardened, you can paint it. Because plaster is very absorbent, paint a thin layer of diluted white liquid glue over the plaster first. Once that's dry, use watered-down paint to paint the leaf and bring out the textures.

Alternatively, you can place the leaf first on the wax paper and pour the plaster over it. To make the leaf curved, mold sand under the leaf to curve it, and then slowly pour the plaster over the top of the leaf.

Finish: Let the cast set out to harden for about a day. Once it's hard, you can clean it with an old toothbrush. Label the back of the cast with relevant information such as the type of print, where you found it, and the date and time. Some naturalists prefer to leave a trace of the substrate so the finer details can be seen. Others prefer to clean the cast thoroughly and then paint it or glaze it. It's entirely up to you.

Try using a flashlight to highlight the crevasses in the animal's footprint. You can then take these casts, which are inverted tracks, and press them into sand to recreate the track to study it as it would naturally look. How many toes does the animal have? Are any claws visible? Was there fur on the foot? Use a tracking guide to figure out what animal made it.

Steps: Sand and Ink Footprint Traps

Make a Sand Trap: A sand trap is best set up at or near your home so you can easily monitor it. Think about where animals are likely to travel through your yard, your alley, or other nearby areas. They typically don't feel safe moving across an exposed lawn but instead move where there's cover.

To make a sand footprint trap, find a tray, like a serving tray or baking sheet— the larger the better. Fill it with bricklayer's or other fine sand, put a container of pet food in the middle, and leave it in a sheltered location away from human activity overnight. You can use a spray bottle with water to even out the surface of the sand before you set it out. Try making casts from the prints!

Make an Ink Trap: Cut a piece of rectangular cardboard wide enough for your expected wildlife to fit through and fold it into three sections to create a triangular tunnel. If you plan to reuse the trap often, cut another piece of cardboard the width and length of the tunnel that can easily slide in and out. If you only want to use it once, then you can prepare the rest directly on the cardboard at the base of the tunnel.

To prepare the base, cut pieces of white paper to the width of the tunnel. Use wide masking tape to affix one piece of paper to each end of the bottom of the tunnel (or to the removable cardboard piece). Be sure to tape both ends of the paper so that it doesn't move. In the empty space between the two pieces of paper, set

> **Tip**
>
> ---
>
> Remember, not all visiting animals will be mammals. You may also get tracks from reptiles and amphibians and even invertebrates. There are very few resources on invertebrate tracks, so it may be difficult to identify them, but slugs, snails, and beetles are among visitors who may leave tracks behind.

the bait. The order of items in the tunnel should be tape, paper, tape, bait, tape, paper, tape.

Mix one part carbon powder with two parts vegetable oil to create a nondrying ink. Use a paintbrush to spread the ink onto the masking tape strips closest to the bait in the middle. If you have enough room, put down another piece of masking tape next to the inked tape and add more ink to make a bigger ink pad for animals to walk over.

Finally, punch holes in the top of the tunnel and thread some twine through them to keep the two ends connected and the tunnel closed. Leave it out overnight and check it in the morning.

Be sure to write your data on the paper when you collect the prints from the tunnel, so you can review your notes later and remember important details.

Find Tardigrades and Other Moss Animals

It is altogether possible that at this very moment, an entire world is living right over your head. Up on the rooftop, in the tiniest clump of moss, is an organism so tough that it can withstand long, hot months without water and an entire season of being frozen. Not only that, but it can survive the vacuum of space and endure doses of radiation that would kill us humans. Some have been dug out of icebergs and revived. They can survive pressure six times that of the deepest part of the ocean in the Mariana Trench. They have become infamous as some of the toughest organisms on earth. They are tardigrades.

Tardigrades have been found living on Antarctica, up on the slopes of the Himalaya, and bathing alongside visitors of

Tardigrade, also known as a water bear and moss piglet

Japanese hot springs. But they also live in the city, unseen on our rooftops, in our lawns, in ponds, and in the tidal zone.

The lives of tardigrades are intrinsically tied to water, but they can survive without it, like when they're shot into space or when rooftop moss dries out in summer. Their secret is called cryptobiosis, a state of suspended animation. When the water dries up, the tardigrade turns into what's called a tun, a state in which their legs and head shrink into their bodies until they look like a small barrel. Their metabolism becomes virtually undetectable, and they are practically indestructible. Tardigrades can stay in this suspended state for years, like something straight out of a science fiction movie. The only thing that will revive them is water.

The question of how long an individual tardigrade can live is a tricky one. While in the cryptobiotic state, they're not actively living despite being alive. So while some tardigrades have successfully been revived after three decades as a tun in Antarctica, their *active* life span ranges closer to only two and a half years at most. And many tardigrade species cannot turn into a tun or survive extreme conditions.

Over a thousand species of tardigrades live their lives plodding around on eight stubby legs, each tipped with four to eight long claws that help them move through moss. Their claw arrangement is a key trait to help identify the different species. Although moss and lichen may not seem like a habitat for aquatic creatures, it is in miniature. Moss is incredibly adept at capturing and storing water among its leaves. The term for this is *limno-terrestrial*, a habitat that alternates between wet and dry.

To call them tiny is an understatement. A 1.5-millimeter-long tardigrade is a giant in its phylum. Most range from 0.5 to 1 millimeter, roughly the thickness of a credit card. Their rotund bodies resemble naked bears or pigs, earning them their endearing common names of "water bear" and "moss piglet." Their faces are utterly alien and yet somehow adorable. Imagine a naked, pudgy, squashed-up bear face with a short, wide folding telescope in place of the snout, eyes, and mouth, and you get the general idea.

Inside their curious mouths are a stylet and buccal tube. This apparatus looks like a weird turkey baster that only comes out when the tardigrade is feeding. The rest of the time it's tucked away inside their mouth. The stylet not only looks like a turkey baster but also functions largely the same way because tardigrades subsist on a liquid diet. But that's not to say they aren't carnivores. Some species feed on bacteria or plants, but others consume animals. A few are even cannibals. Regardless of what they eat, they all stab the needlelike stylet into the cells of their food source and suck the fluids out. When they're done, they pull the stylet back inside their mouths.

Under a microscope, a tardigrade has no secrets. With the backlight shining through the tiny tardigrade's transparent body, you can see their stylet even when they're not feeding. Not only that, but you can also see if they are carrying eggs. You can study their claws and watch as they scramble for purchase. Naturalists can also study the tardigrade's shed skin, which they leave behind when they molt. You may even find eggs inside those shed skins where the female laid them when molting.

Materials

Microscope: The only way to view tardigrades is with a microscope, so this is one of the only projects in the book for which you'll need something relatively expensive. A standard dissecting microscope costs around a hundred dollars, but you can find some for less and some for far more. There are also newer options, such as handheld origami-style ones and laser-cut wooden models that cost less.

Tardigrades are tiny, but you can see them fairly well with a basic dissecting microscope with a top and bottom light. The bottom light is very helpful because tardigrades are transparent and you can often find eggs in them. Having a compound microscope with higher magnification isn't necessary, but it will allow you to view them even larger and see more details.

Petri Dish: The next thing you'll need is a petri dish—or really, any small, clear container with a flat bottom. Glass dishes you find at a thrift store or in your kitchen cabinet, or plastic containers from your recycle bin, can all work. You may want to have a few on hand for multiple samples.

Rainwater: You'll need some water to soak the moss, and rainwater is best because it generally lacks the hard-water minerals, additives, and chemicals found in tap water. If rainwater isn't available, tap water will be fine.

Pipette (optional): You'll only need this if you want to look at the tardigrades under a compound microscope. It's very hard to pick up a tardigrade and put it on a slide, but a fine-tip pipette works well to transfer them.

Desk Light (optional): Having a secondary light with a bendable arm can help illuminate tardigrades from the side since many are white or clear and can be tricky to see with only the microscope light.

Steps

Collect Moss: A naturalist may find tardigrades in mosses or lichens growing on their own driveway or in a patch on the sidewalk. These are great places for experimentation, because some moss may rarely have any tardigrades, whereas other moss can be practically swimming with them. Look for moss that grows in compact tufts, like *Bryum argenteum* or *Ceratodon purpureus*. Try collecting a few clumps of various types of mosses from different locations, like a lawn, driveway, fence, rooftop, and some rocks. You don't need large samples—clumps the size of a quarter or half dollar will work. Try not to leave too much soil or other debris on the clumps because it will cloud the water, making the tardigrades harder to find. If you collect moss from different locations and want to document what you find, put each clump in individual dishes to keep the results separate.

> ## Tip
>
> Instead of disposing of the water used for soaking, save it in another dish and put it under the microscope. You may find tardigrades in that water too.

Soak: Put the clumps of moss into your dish and fill it up with rainwater. Let the moss sit in the water overnight or at least for a few hours to soak it up. If you've had a recent rainstorm and the moss is already soaking wet, you may be able to skip this step.

Drain: Remove the moss and set it aside in another dish, or just hold it, while you dump out the remaining rainwater from the petri dish.

Squeeze: Hold the moss clump in your hand and squeeze it over the empty dish, just like squeezing out a sponge. You should get a good amount of water in your dish. Set the moss aside. If you're using different types of moss or moss from different places, squeeze the clumps into their own dishes to keep them separate.

Search: Put the dish under the microscope and look. If you're lucky, you'll see tiny tardigrades moving around, but you may have to search for some time. If you're looking through a dissecting microscope, they will be small, but they move often. They come in different colors, from white to pinkish or gray. Try flipping between the scope's backlight and top light and look for eight stubby legs, each with a set of long, curved claws. The two back legs will look like they're coming out of their bum. Tardigrades are chubby, and although they stretch out, they don't change shape much.

> ## Tip
>
> If you have a flexible lamp, angle it down at the level of the dish and shine it in from the side. Because tardigrades are often transparent or white, overhead lighting can wash them out. Shining light on them from the side can help reveal their details. Also try flipping between the overhead light, backlight, and side light.

Document: Once you find tardigrades or other moss animals, make notes or sketches in your nature journal. Observe their behavior and study the way they move when they're not in their natural element of moss. Turn on the backlight to study their anatomy and see if you can find any with eggs. Look for the shrunken, barrel-shaped tuns, and if you find any, try watching them to see if they revive in the water. Make notes in your journal of the type of moss, the date, how many tardigrades you found, and any other relevant information.

Moss Invertebrates

Over five hundred million years ago in the Cambrian period, tardigrades lived alongside trilobites and other now extinct invertebrates. While trilobites no longer exist, you will likely encounter a number of other interesting invertebrates in your hunt for water bears. Some are water breathing, like rotifers and nematodes, and live in the water film on moss. Others breathe air, like mites and springtails, and live mostly on the surface. In addition to the invertebrates described below, you may also encounter some truly tiny organisms, like ciliates, which are protozoans, and diatoms, which, although they move around, are actually algae!

BDELLOID ROTIFERS (BDELLOIDEA)

These microscopic animals are similar to tardigrades in many respects, including the ability to survive in space and other extreme conditions. But rotifers possess a different survival trick. Once their moss habitat begins to dry out, they secrete a mucus that hardens around them, creating a sort of chamber where they stay until moisture returns.

You are likely to encounter rotifers because they are abundant in the same moss habitats as tardigrades, sometimes more reliably so. However, they look and move differently. Rotifers are roughly shaped like a champagne flute, wider at the head and tapered at the other end. Depending on their position, they can sometimes look more goblet shaped. Rotifers are remarkably stretchy, thanks to a retractable foot that acts like a telescope, and when they want to travel, they reach out with one end and pull their body after, almost like an inchworm. They can be ball-like one second and wormlike the next, and they move quicker than tardigrades. When they settle down, you may see their wheels moving. Rotifers suck in water using their hairlike appendages called cilia, which look like fast-spinning wheels, to filter algae, yeast, and bacteria out of the surrounding water to consume. In fact, their name means "wheel animals." All bdelloid rotifers are female. They reproduce by parthenogenesis (a type of asexual reproduction) and therefore have no need for males.

NEMATODES (NEMATODA)

Chances are very high you'll encounter nematodes, which are small compared to earthworms but huge compared to tardigrades. These worms are abundant just about everywhere on the planet, including in moss alongside tardigrades. In fact, out of every five animals on this planet, four are nematodes. Many, if not most, animal species have their own personal parasitic nematodes, but nematodes also feed on plants, bacteria, and fungi. Some are predators, and others are scavenging opportunists that eat whatever they can find.

While tardigrades plod and rotifers inch along, nematodes thrash about wildly and are impossible to overlook. Nematodes are more likely to be found in consistently wet moss, but they're not without defenses if the moss begins to dry up. Nematodes migrate within the moss, moving deeper down during the day to avoid the dried outer edges. But like tardigrades and rotifers, nematodes can tolerate some degree of desiccation, slowly becoming dehydrated.

MITES (ACARIFORMES)

Some mite species are aquatic, like the ones featured in the pond-dipping project (Project 2), but others live in the liminal zone between wet and dry environments.

Few mite species actually live in moss, but many spend some of their time in or on it. The mites you might see while searching for tardigrades will either be on the surface of the water, or sadly, drowned at the bottom.

Some mite species are bright red and are easy to spot. The color comes from carotenoids, which are believed to act as a type of sunblock, protecting the mites from UV light. But many species, like those of the Oribatid order, are not as glamorous, being brown or other earth tones.

Mites are arachnids and have eight legs, except when they are larvae and have only six. Their thorax and abdomen are fused together, and there's no separation, unlike spiders. Their bodies are generally quite plump, and their legs are scrawny. Moss mites move slowly and feed directly on both the leaves and capsules of moss. Some species feed on dead plant matter, and some eat fungi and algae. Other mites are parasitic and are looking for other invertebrates that live in moss. Mites have few defenses, so at times they play dead like opossums.

SPRINGTAILS (COLLEMBOLA)

It's possible that springtails are older than the moss they live on and may have been jumping around the world before mosses arrived. Regardless, moss has long been a home—but not the only one—to these invertebrates.

Like other moss dwellers, springtails also prefer to keep themselves wet. But since they breathe air, they live on the aquatic edge of moss like mites do. They're not as adept as tardigrades and rotifers at tolerating desiccation, but they do have a few tricks, like a grooming fluid they can produce to keep their heads moist when the moss dries out. It's almost like a reversed diving helmet that keeps the water in instead of out. They spend their time crawling around in the moss, feeding on fungi, mating, and laying eggs. The miniature, leafy jungle of the moss also serves as protection from predators.

Springtails can vary in appearance. Some are long and slender with six legs and long antennae, and others have extremely round abdo-mens. They can be a range of colors, from pink to red, yellow, and white. Some are iridescent and shine under the microscope's light. Since they are not aquatic, they won't be found in the water sample but floating on top of it. And possibly trying to spring away. Many species have phenomenal jumping abilities, relative to a human jumping over the Eiffel Tower.

For more about springtails, see Project 9, to learn how to make a Berlese funnel.

Find Your Face Mites

SEASONS:
All

STUDY TOPICS:
Face mites, parasites, microscopy

You're never alone when you have face mites, and nearly everyone has them. They're one of our oldest, most constant companions. Humans have had face mites ever since *Homo sapiens* evolved. An individual person can have anywhere from hundreds of face mites living with them to thousands, possibly even millions.

We're not born with face mites, and how humans first acquired them is still a mystery. Like our genes, scientists believe they are passed to children during those intimate first moments when parents embrace their infants, and from then on, they continue to go with us everywhere—even into space and the depths of the ocean.

Finding them can be hard, however. For one, they're tiny. At a mere 0.3 millimeter long, mites generally live undetected by their

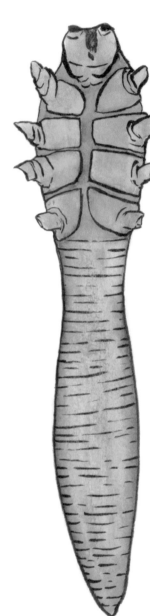

human hosts. A whole lot of face mites could fit comfortably on a 2-millimeter pinhead.

Also, even though we know everyone has them, scientists have found them on only about 14 percent of people they have sampled. Mites can be unevenly distributed on faces. You might have them only on one cheek, or only on one side of your nose. But testing has shown that mite DNA is found even when mites are not, proving that virtually everyone has face mites.

Before you start getting itchy and go scrub your face, you can't get rid of face mites. DNA analysis suggests you don't only get them once, but it's likely you have been colonized multiple times throughout your life by face mites. Not only do parents share them with children but adults share them with one another through close contact and through shared textiles such as pillows and towels. Even with medical treatment, they come back.

Face mites are in the *Demodex* genus, and as the name suggests, they live on your face, where they spend their days residing in pores snuggled up with your hair follicles. Hidden away face-first in their cozy home, they consume the oil produced at the base of our pores, called sebum, which is in the hair follicles and keeps our face from drying out. The greasiest areas of a person's face, usually the forehead and nose, are the best places to find face mites.

Demodex are considered commensal organisms rather than parasites, because their relationship with humans benefits one but does not impact or do harm to the other. For a small number of people, face mites can cause problems, usually when a person's immune system is compromised in some way, like when they are going through chemotherapy or using hydrocortisone cream. In these circumstances, face mites can become too numerous, causing a rare condition

Demodex folliculorum

face mite

called demodicosis, which some scientists call the "*Demodex* frost" because of the visible white sheen that develops on someone's face.

But what do face mites look like? They are only visible with a microscope. Imagine a *Tyrannosaurus rex* but more wormlike and with eight stubby arms, four lining each side. Instead of back legs, face mites have a long, thick tail. And their body is semitranslucent.

While you are fast asleep at night, your face mites crawl out of their pore homes and onto the surface of your face. You see, a single face mite lives for only two weeks, and to keep a population stable, they need to reproduce. While you're dreaming, they are traipsing across your face, looking for a partner.

Even though it happens right under our noses, we know almost nothing about how face mites mate. When they return to their individual pores, the female mites lay their eggs, which are up to *a third to half* the size of the adult mite's body. They likely don't lay very many, and possibly only one at a time.

Until very recently it was believed that because they have such a short life span, face mites did not have an anus and did not expel any waste during their lives—debunked by a recent study. Sometimes scientific research alters what we think we know.

We're not the only mammal to have these companions. They have been found on nearly every single other mammal scientists have tested. Different mammals have their own species of face mites, and humans have not one, but two different, only distantly related species of face mites. *Demodex folliculorum* is long and thin and lives near the pore's surface, and *Demodex brevis* is short and squat and prefers to live deeper in the pore. Some people only have one or the other, while other lucky individuals have both. For reasons not yet known, people living in different geographic regions have different species and lineages. Mites from China are genetically different from the mites of the United States, for example. Such distinctions may help reveal details about the earliest human migrations, telling us whether different modern populations are more closely or distantly related.

So far scientists have distinguished four different lineages, but there are many more populations to sample, so that number is likely to grow over time. While *D. folliculorum* is pretty similar from person to person, *D. brevis* is not. And because the two species of human face mites are not closely related, the current theory is that we picked them up in different ways. *D. brevis* is more closely related to the face mites that dogs have, so it's possible dogs shared them with humans.

Scientists are just now starting to grasp the long and very intimate history we humans have had with our face mites. No matter how close you are to your kids, partner, or even dogs, you're even closer to your mites. And over your lifetime, you will host hundreds of generations of face mites. It turns out that we each have our own genealogy record living on our face. And there's still a great deal we don't know about these mites. You can't wash them away, but you can learn more about them and appreciate your peaceful companions with this project.

Materials

The tools you'll need to find face mites are few but specialized. There are two methods for observing face mites that you can try, depending on your materials and comfort level. The first is the simplest and uses tape. The second uses superglue on microscope slides.

Microscope: Because face mites are so small, you will need a microscope to look for them. A simple compound microscope will be suitable.

Slides: You'll need basic microscope glass slides to view the mites. If you use the superglue method, you can use each slide only once, so you'll need a box to test multiple times.

Tape: The tape method uses basic cellophane tape. Clear tape works better for viewing under the microscope, but nonclear types work fine. You can also experiment with over-the-counter pore strips or packing tape.

Superglue: The second method uses basic superglue, but not all superglues are created equal. Standardized superglue is what scientists use. Avoid anything colored, rather than clear, because it can obscure viewing. Also avoid glue that is too liquid-like—it can run off the slide and not bond. You may have to experiment because some glue won't bond with the glass, while others may obscure the slide's contents under the microscope.

Steps: Tape Method

This technique captures mites that come out at night to mate or migrate across your face. It's easy, painless, and risk-free.

Tip

Face mites desiccate quickly (within about five minutes) once removed from the human face, so look at the slide immediately.

Place Tape: Place a strip of clear tape, one to two inches long, on your face before bedtime. You're likely to have the most success on your forehead, cheeks, or across the bridge of your nose. Leave the tape on your face overnight. You can also try applying tape during the day for a shorter period of time, about a half hour. To maximize your chances of finding mites, try placing several pieces of tape on your face at once.

Remove Tape: Peel the tape from your face and place it on a new microscope slide, sticky side down.

View: Place the slide glass side up, tape side down, on your microscope and look. Use between 40× and 100× magnification. With this method, the mites will be near the surface of the slide, so focus on that layer. Search carefully because they can be hard to spot. Once you find one, note the appearance, and even make a sketch, to help determine which *Demodex* species you have found. Keep searching and note how many you find. Consider making notes in your nature journal of your process and results.

Steps: Superglue Method

This technique is called the "standardized skin surface biopsy" and is the technique that scientists commonly use to extract face mites.

Apply Glue: Apply a drop of superglue to the center of a glass microscope slide. Don't put more than a drop because if there's too much, you'll leave the glue behind on your face instead of on the slide.

Place the Slide: Immediately place the slide, glue side down, against your

Warning

Use this method at your own risk. Studies indicate superglue shouldn't damage your skin, but use extreme caution, especially if you have dry or sensitive skin, irritation, or other skin problems. The fumes can also be irritating to eyes, so close your eyes if you're placing glue near them.

face. The best areas are your forehead or cheek. Leave the slide in place for a minute to give the glue time to dry.

Remove the Slide: Very gently peel the slide off your face.

View: Place the slide, glue side up, under the microscope starting with 40× magnification and moving up to 100×. With this method, the mites will be found at the end of the follicles, farthest away from the skin on the slide, so focus on the tips higher up. They aren't always obvious, so examine the slide carefully and methodically. If you see one, note its anatomy, sketch it, or try taking a photo with your phone through the eyepiece.

Document: Make notes in your field journal of your specific process and what, if anything, you found.

Repeat: By repeating the procedure in the same place on your face, you can extract deeper levels on each subsequent pass and possibly collect mites that may live deeper down in your follicles.

The Mites

The two types of face mites that live on human faces are easy to distinguish.

DEMODEX FOLLICULORUM

This species tends to live higher up in pores, residing above the sebaceous gland. They are larger and longer bodied than *D. brevis*, with a round bottom. There are often several individuals per hair follicle. To feed, face mites clamp on to their food source, and this species has an organ called a palpus on their face with seven claws.

DEMODEX BREVIS

Stubbier and smaller than *D. folliculorum*, and with a more pointed bottom, this species lives deeper in the pores, embedded in the sebaceous gland. They live in fairly light densities, only one or a handful per gland. The palpus of this species has five claws.

Go Sugaring for Moths

By the time butterflies arrived on the scene, moths were already there, waiting for them. Moths evolved long before their day-flying counterparts, and they are massively more diverse.

Moths and butterflies make up the order Lepidoptera, and together they number around 180,000 species documented so far. (Many moths have not yet been described, so that number will surely increase.) Of that number, moths account for 165,000 described species, or 90 *percent* of all species in Lepidoptera. Despite that fact, butterflies get most of the glory, while moths are frequently relegated to "pest" status, even though only a tiny handful are true pests to humans. The rest are harmless, and many

Underwing moths

are highly beneficial as pollinators or an essential source of food for wildlife like bats or nesting birds.

Living in the darkness is one of the main reasons moths have such a negative reputation. They are mysterious, flying around in the night largely unseen, occasionally banging against a lit window and frightening someone inside. Humans are generally afraid of the night and what lives in the darkness. Living in cities that are bright even at night, most of us never know true darkness. Few of us are comfortable in the dark or ever venture out into areas unlit by streetlights. Humans generally fear what we cannot see, and in our evolutionary past, we had a lot to fear.

However, in more modern times, a naturalist's curiosity can outweigh their fear, and naturalists have long been curious about what goes on at night outside their bright homes. Over the generations, naturalists have developed many methods for investigating the nocturnal world, including "mothing."

Mothing became easy with the advent of electricity. A flip of a switch could attract moths and shine a light on the mysteries of their nocturnal world. Naturalists have since perfected this technique with specialized and expensive light trap equipment designed to attract a wide range of moths, as well as simpler methods like hanging a light bulb over a white sheet. Today these methods are widely used by "moth-ers" and scientists alike. A naturalist at home can also use these methods, the simplest one being leaving a porch light on.

But there's an even older method, devised before electricity was common, and perhaps even before it was invented. Sugaring for moths is such a common and popular technique of attracting moths that its origins warranted a scientific paper. That's partly because multiple people fought over the distinction of having invented the method of sugaring for moths, which involves painting a sugar mix onto trees. However, the paper's authors found instances that went back even before the disputes and recount Victorian naturalists smearing honey, sugar, and other sweet substances on surfaces outdoors to attract moths. This method surely came about from naturalists' documented observations of moths feeding on rotting fruit, such as blackberry and yew, and being attracted to tree sap. Regardless of its origins, it's been a popular method for the last two hundred years among naturalists, moth-ers, and scientists.

There is no one-size-fits-all approach for attracting the many families of moths. While light is an excellent way to draw certain groups, sugaring is far superior when it comes to some in the Noctuidae family, notably those in the *Catocala* genus, more commonly known as underwings. Because sugaring for moths has been so widely used over such a long time, there are a multitude of recipes and tips. Each moth-er seems to prefer their own combination, and a lot depends on geographical location, time of year, habitat, and other factors.

> ## Warning
>
> If you live in an area with bears, use caution. The sugar mixture may attract them—you certainly don't want a surprise visit from a bear in the dark.

There are two variations of sugaring for moths. One is called the wine rope method and consists of soaking a rope in a mixture of wine and sugar and then hanging it outside. The other, more common sugaring method includes cooking some ingredients and painting the mixture on a tree or fence post.

Sugaring for moths is possible year-round in most parts of the world, but generally, the best time for sugaring in the Northern Hemisphere is mid-July through the end of summer. The ideal conditions are warm, humid nights with a light wind and little to no moonlight. A gentle breeze helps carry the scent of the sugar mixture to hungry moths. Heavy wind, however, usually prevents them from flying, as do bright moonlight and rain. Autumn and winter are also excellent seasons to try because it's a time when easy sources of nectar, like flowers, are waning and moths are desperate for food.

You don't necessarily need to make a mixture to sugar for moths. When you're in the woods, keep an eye out for active sapsucker holes or tree sap flowing from broken branches or a damaged trunk, then visit those places after dusk. If you live in an area where maple tree sap is harvested, visit at night when the sap starts flowing, usually in February and March. Or note locations with overly ripe and rotting fruit, like blackberries. Moths likely visit these locations and can be observed during a night walk.

As for all naturalist activities, there is likely to be a period of trial and error. Despite your best efforts, you will not be successful every night. Every naturalist knows that our fellow creatures sometimes behave inexplicably, leaving us baffled. If that happens, keep trying.

Materials & Ingredients

This project reads more like a recipe, but it is a lot looser and more customizable than your typical recipe in a cookbook with specific measurements.

SUGARING METHOD

The basics are sugar (brown works better, but white can work too), molasses, stale beer, and overly ripe fruit. The most common fruit is a banana, but moth-ers have successfully tested a wide variety of fruit including mangoes, pears, and peaches, so experiment with different combinations. Some moth-ers swear by overly ripe watermelon and smash it directly onto the tree by itself. Stale soda can be used in place of beer. Many moth-ers add a little rum. Other possible ingredients are vanilla, yeast, sweet liqueur, and maple syrup.

Paintbrush: Find a paintbrush that is several inches wide. You'll use it for a sticky, stinky concoction, so don't use your best paintbrushes.

Containers: You'll need a pot to make the mixture and a lidded container to store it. Make sure the container has an opening wide enough for your paintbrush to fit into. You may also want containers to catch moths in to observe, photograph, or draw. Any small clear container can work.

> ## Tip
>
> If you have a *Catocala* moth at your sugar and you want to see the underwings to identify it or take a photo, you can gently touch its abdomen or backside. The moth will likely spread its wings, revealing its colorful hindwings.

Light and Filter: You can use a simple flashlight, or even better to keep your hands free, a headlamp, to check the sugar mix for moths at night. Cover the light with a red filter, like a piece of cellophane, or use a red light, a feature of some headlamps, to avoid scaring away the moths.

Sponge (optional): You can also use a sponge hung from a tree with string if you don't want to temporarily stain your tree or fence post.

Gloves: Rubber gloves can help keep your hands clean.

Camera: If you want to document what you see, bring a camera with a flash. Phone cameras work as well.

WINE ROPE METHOD

You'll need a bottle of cheap red wine and a couple cups of sugar.

Rope: You'll want a thick rope made out of a material like cotton that can absorb the mixture, about the thickness and texture of a mophead. Alternatively, you can use strips of fabric.

Steps: Sugaring Method

Scout a Location: Before night falls, decide where you want to sugar for moths. It's a location you'll visit in the dark, so find somewhere safe and away from dangers like cliffs, roads, or other hazards that are harder to see at night. Backyards work just fine, but avoid areas that are brightly lit. If you can go farther out, trees at the edge of a woods are ideal. You can also paint trees along trails. But be sure to read and follow the rules of where you are, including park and public lands hours. Once you have some experience, consider talking to park officials to create a moth viewing event.

Create the Mixture: Put the beer or soda into a pot on the stove and simmer, but don't boil it, for five minutes. Add the sugar and molasses and let it simmer for another five minutes or until it's thoroughly dissolved. Smash up the fruit and mix it in. There are no hard-and-fast measurements, but try roughly one can of beer, two cups of sugar, and one piece of fruit, then add molasses until it's thick.

> ### Tip
> If the mixture isn't thick enough, try adding more molasses.

The mixture should be gelatinous; the consistency is more important than proper proportions. It should be fairly thick—thin enough to paint on, yet not so thin it runs down the tree trunk. The main goal is to make a mixture that is sweet with a strong odor.

Allow the mixture to cool to room temperature. Add a little rum, if you are using it, just before painting it onto the trees.

Decant: Pour your mixture into a container.

Ferment (optional): You can let your mixture ferment at room temperature for a few days before applying it, which some moth-ers believe makes it even more appealing to moths. Place it in a container *without* an airtight lid (otherwise, it might explode). A paper towel held on with a rubber band works well.

Paint: Head to your location to apply the mixture around or just after dusk. Painting too early can attract diurnal insects like wasps, and the mixture can also dry out too quickly. Use the paintbrush to paint the sugar mixture on a tree trunk or fence post about a square foot in size and at least four or five feet off the ground, preferably at eye level. Avoid placing the mixture where someone might lean against it. Try not to paint over lichens and mosses, which could damage them. Also avoid letting the sugar mix drip, to reduce the likelihood of other insects like ants finding your mixture.

Observe: The first couple of hours after dusk are the most productive time for moth activity. Keep your voice low and walk quietly to avoid scaring them. Use a red cover over your light, pointing it down and away from the spot as you approach. Most moths will be camouflaged, so look closely at the trunk or try looking from an angle so you can see their shadows. Note their camouflage and see if they blend in with that particular type of tree trunk. Look for their proboscis, the insect version of a tongue, which curls in front of their face when not in use and can be quite long in some species when they're actively feeding.

If you want to study a moth more closely, draw it, or share it with a friend, you can carefully catch it. Entomological supply stores sell specialized containers online for observing moths, or you can reuse a plastic or glass container. Old tape cassette containers also work wonderfully to hold the moth's wings flat so they don't flail around and get hurt. Be sure to release the moth when you're finished observing it.

> ## Tip
>
> If ants become a problem, you may have to try another tree or fence post elsewhere. If your options are limited and the ants are keeping the moths away, try painting a ring around the tree with the same mixture below the square for moths to keep the ants busy down there.

Identify: Moth identification books are scant compared to what's published about butterflies, but there are still good resources available. Many moths can't be identified to species without a lot of experience (and sometimes a microscope), so consider learning the families at first, and if you develop an interest in moths, you can always dive deeper. There are helpful silhouette guides in moth and butterfly field guides and online. Check the "Moths" section in Resources for some recommendations. You can also share your photos to online moth groups or insect identification websites like BugGuide.

Repeat: The mixture will only be effective for one night, possibly two. You'll need to reapply it if you want to observe on another night. However, don't sugar in the same space more than a couple nights in a row. Give the moths time to disperse, and avoid creating a predictable, convenient banquet for any of their predators.

Document: Record your observations with photos and notes and submit them to community science projects like National Moth Week, which takes place annually in July, or one of the many other moth community science projects around the world. These records can help document species ranges, flying times, and other important information. You can also put your observations in your nature journal.

Steps: Wine Rope Method

Make the Mixture: Heat the wine in a pot on the stove and add the sugar. Once it's dissolved, let it cool to room temperature. Rough proportions are one bottle of wine to two cups of sugar.

Soak the Rope: Place the rope into the pot and let it soak for ten minutes. Put on rubber gloves. Remove it from the container; let it drip and give it a shake, but don't squeeze it out.

Place the Rope: Hang the wine-soaked rope on a fence or over tree branches. See steps above for the rest.

Moth Life: Underwings (Catocala)

The moths most likely to visit your sugar mixture are those in the *Catocala* genus, also known as underwings. The name *Catocala* comes from a Greek root and means "beautiful below," a fitting name for moths with colorful hindwings of red, pink, orange, or yellow. All moths have four wings, but unlike butterflies, when moths rest, their forewings generally cover and hide their hindwings. When underwings are at rest, their mottled-brown forewings hide their very colorful hindwings. Those are only visible when the moths are active.

The bright hindwings are thought to be a deterrent, ready to flash open to startle any predator. However, studying the less colorful forewings reveals a subtle yet intricate beauty of swirling, marbled patterns in every shade of brown and gray imaginable. Brown doesn't come close to describing the shades found on underwings. Umber, chestnut, mahogany, chocolate, taupe, ochre, russet, and sepia barely touch on the many tones of underwing moths. Some even have a silver sheen. The patterns and colors of the forewings mimic the bark of trees on which these woodland moths rest during the day. Their habitats vary from oak woodlands to mixed hardwood-conifer forests.

Larvae of underwing moths generally feed on the leaves of trees such as oaks, aspen, willows, and walnuts, among others. The mostly hairless caterpillars are excellent branch mimics—their brown, green, and gray speckled bodies don't so much blend in as look like the wood of smaller branches. There are roughly 230 species of *Catocala* worldwide, and while moths vary dramatically in size, this genus tends to be large and noticeable, with wingspans from 20 to almost 100 millimeters.

Interestingly, Carl Linnaeus, the father of taxonomy, gave many of the *Catocala* species names with a female marriage theme. So you may encounter "the bride underwing," "the old wife," "the widow," or even "the once-married underwing."

Make a Berlese Funnel

SEASONS:
Spring, autumn

STUDY TOPICS:
Leaf litter, soil, and ground invertebrates

A healthy forest begins with its floor. In autumn, leaves fall by the thousands, piling up on the soil. Wind breaks moss- and lichen-covered branches off trees, sending them down into the debris. All manner of other organic material ends up on the forest floor as well. While the common term is *leaf litter*, it is anything but garbage. All that dead matter plays a number of important roles in the health of the forest and the lives of a huge diversity of forest wildlife species.

Deer, raccoon, bear, and fox are what many people would say if you asked them to name traditional woodland animals. But under the feet of those scant few mammals lives a far more populous group of creatures, most of which are rarely seen or considered by human wanderers.

Life on the forest floor

Invertebrates, as you may have discovered from other projects, live nearly every-where in the world, and the forest floor is no exception. In fact, forest invertebrates far outnumber the vertebrates like mammals, birds, amphibians, and reptiles, and without them, the forest ecosystem would likely collapse.

The vast majority of forest invertebrates live in the upper layers of soil and debris on the forest floor, among the fallen leaves and branches. These layers form a sort of magical sandwich that houses a large diversity of wildlife, serves as the base of the food web, and keeps the soil and nearby plants healthy. And on top of all that, the leaf litter and soil layers offer a wonderland of exploration for any naturalist.

Most residential yards lack this invaluable layer because modern humans have an incessant (yet largely unnecessary) need to tidy everything. When homeowners rake away fallen leaves each autumn, sticking them in bags to be hauled away or mulch-ing them into small pieces, they remove invaluable shelter for a host of invertebrates, and throw out the overwintering butterflies, bees, and other garden friends.

The best place for black-and-orange striped woolly bear caterpillars to spend the winter is a pile of leaves. A lone overwintering queen bumble bee, who will face the monumental task of beginning a brand-new colony in the spring, rests just under the soil, snug beneath the debris on the forest floor. Insects overwinter in different phases of their life cycle, some as eggs, others as pupa, and a few as adults. Leaves and other debris act like a cozy quilt, keeping them warm and sheltering them through the cold seasons.

But not all invertebrates are resting. Many are active on the forest floor throughout the year, where they serve as the base of the food web—a welcome meal for birds, turtles, chipmunks, and many other animals. Springtails and mites, snails and slugs, millipedes and centipedes, spiders, beetles, and many other even smaller animals live hidden away in the woodlands.

Without decomposer invertebrates, along with bacteria and fungi, fallen leaves would lie on the forest floor forever. Decomposers break down leaves, seeds, wood, and every other type of organic matter. Thanks to them, leaves that protected invertebrates and slumbering plants in winter take on a new role as fertilizer, releasing valuable organic matter back into the soil an annual and ongoing process of nutrient cycling.

When you walk in the woods, at your feet is a biodiversity as great as any exotic landscape. Some invertebrates live their entire lives in an area the size of a yard, while others never travel farther than an area the size of a leaf.

For many years naturalists and scientists painstakingly sorted invertebrates out of leaf litter and soil by hand, until an entomologist came up with a simple, more efficient method. In the 1880s, Antonio Berlese developed a new way to have the invertebrates perform nearly all the work. He knew that leaf and soil invertebrates aren't fond of light, because the heat dries out their home. Tiny creatures can die remarkably quickly when their bodies dry out, so when light shines on their shelter, they move away from it. Taking advantage of this light-avoidance behavior, Berlese built a simple tool using a funnel, catchment container, and heat source. The invertebrates in the funnel move away from the source of heat, deeper into the funnel until they fall into the catchment container. This tool, called the Berlese (*bur-LAY-zee*) funnel, is simple to make with materials readily available in most households.

Materials

Shovel: While you can use your hands to scoop up leaf litter, a small hand shovel will save you from getting your hands dirty and allow you to dig into the top layer of soil. A fold-up metal shovel is a good general tool for a naturalist's field bag and worth buying, but any shovel you have on hand will do.

Bag: You'll need something big enough to carry your collected leaf litter and soil home. A reused plastic bag or even a plastic container will keep the mess together and prevent the sample from drying out too quickly.

Funnel: The size of funnel is entirely up to you. The bigger the funnel, the bigger the sample size you can filter through, so ideally you'll use something that can hold at least a liter or so.

You can purchase a funnel or make one out of a liter soda bottle or milk jug. If you're making your own funnel, there are two approaches. To make a larger funnel, use scissors or a knife to cut the bottom off a soda bottle or milk jug. Place the bottle upside down, with the top inverted into a mason jar or other container (see full description below).

To make a smaller funnel, you can use the entire milk jug or soda bottle and eliminate the need for a separate container. Cut the jug or bottle in half. Take the top half, the one with the spout, and flip it upside down to create the funnel. Place it inside the bottom half of the jug or bottle. If it slips through to the bottom, tape the two halves together.

Container: If you are using a regular funnel or the larger funnel made out of a milk jug or soda bottle, find a stable container that it will fit on. It needs to be sturdy enough that the funnel with leaf litter won't fall over. Mason jars usually work well for this. Depending on the size of your container, you may also want a smaller dish to place under the spout to catch tiny inverts so they stay contained. A petri dish or any other small, shallow container will work.

Screen: A piece of hardware cloth or window screen will hold the debris in the funnel and allow the invertebrates to fall through. If you use something with a finer mesh, like a window screen, cut a few slits in it for the larger invertebrates to fit through and fall into the container.

Lamp: Use a simple desk lamp with a 20- to 40-watt incandescent bulb, preferably the kind with a bendable neck, to shine down into the funnel. The smaller the funnel, the smaller the wattage you need. Anything higher than 40 watts may kill the invertebrates.

Distilled Water and Paper Towels: To keep the invertebrates from drying out and dying, you'll want to provide moisture, and distilled water is the gentlest type. Paper towels will help retain the moisture in the container.

Microscope or Hand Loupe: Many of the leaf litter invertebrates are very small, and you'll see them best with a dissecting microscope, but you can see a great deal with a hand loupe.

Steps

Find Leaf Litter: You can sample nearly anywhere there are trees. It can be as simple as under a single yard tree or in a woodland or forest. Try sampling multiple locations and comparing results. Autumn is the best time of year for leaf litter, but forests are full of debris all year long. Try sampling in different seasons.

Collect: Use your shovel to scoop up enough leaf litter and other debris to fit into your funnel. If you'd like, also scoop out the top layer of soil. Place the leaf litter in a plastic bag or container to preserve its moisture.

Assemble Materials: You'll want to keep the invertebrates alive when they fall through the funnel, so moisture is vital. Soak a paper towel in distilled water, and place it in the bottom of your collection container or small petri dish. That way, the invertebrates will have a damp place to rest. Set the funnel on top of the container.

Cut or bend the hardware cloth or screen so it nestles inside the funnel, close to the spout, and serves as a platform for the leaf litter to rest on. If the screen is stiff, it can simply sit in the funnel, but if you are using standard window screen or something less sturdy, you may need to tape it inside the funnel. If your screen is flimsy, you can use a rubber band to secure it to the spout.

> **Tip**
>
> If your setup is wobbly, tape a stick to the outside of the container and funnel to stabilize it.

Place the leaf litter in the funnel and arrange the lamp so the bulb is a few inches above the litter.

Note

When scientists use this method, they nearly always put alcohol in the catchment container to kill the invertebrates. The version in this book preserves the lives of the invertebrates because, as naturalists, it's valuable to study the living animals.

If your funnel and container are clear or transparent, put the project in a closet or other dark place, or cover it in black paper or tape so the main source of light comes from the lamp.

Wait: Let the funnel sit undisturbed as it dries out. This process can take up to three days, but you can remove the funnel temporarily to look at the collected invertebrates each day. Add more distilled water if the paper towel dries out. Remove the funnel carefully so soil doesn't fall through the screen.

Observe and Document: You can observe the living invertebrates under a dissecting microscope. Put a lid on the petri or small dish to retain the moisture and prevent the invertebrates from escaping. A fine paintbrush makes a good tool to pick up or move tiny animals. Make notes, sketches, and other observations in your nature journal. Once you are finished, release the invertebrates back into the outdoor leaf litter. If you'd like to observe them over a longer period of time, see Project 12: Make a Habitat Terrarium.

Leaf Litter Invertebrates

There are more organisms living in the top layers of a forest floor than there are terrestrial organisms in the rainforest. In addition to invertebrates, there are also bacteria, fungi, and algae, among other organisms, but a lot are creatures without spines. They cover a huge diversity of orders from worms to springtails and come in a tremendous range of sizes. Some are microscopic, like tardigrades and rotifers. Others are larger and more familiar to most humans, like earthworms and slugs. A naturalist may encounter weevils or the larvae of lacewings, flies, sawflies, and moths in the leaf litter. You might recognize forest floor invertebrates from other projects, like mites and springtails, and woodlice, spiders, beetles, and earwigs.

Entire books have been written on the wealth of diversity found under our feet. Here are a few of the most common ones you may be lucky enough to encounter.

Investigate a Leaf Litter Sandwich

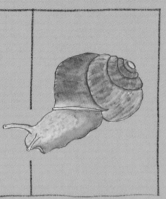

You may wish to investigate the soil and debris of the leaf litter layer by layer at your leisure, which requires only simple tools. First, find a biscuit cutter that's at least three inches deep and wide and a piece of stiff cardboard.

Select an area of forest floor and press the biscuit cutter into the ground. Clear the area around it with your fingers until you reach the bottom of the cutter. Then slide the cardboard under it. You may need to slice the soil first with a knife, depending on the composition.

Pick up the whole leaf litter "sandwich" and place it in a bag. Return home, lay the entire sandwich on a paper towel or in a large petri dish, place on a dissecting microscope, and start to pick it apart, layer by layer. If you don't have a microscope, use a magnifying glass or hand loupe instead. Use your journal to make notes about your discoveries.

PSEUDOSCORPION (PSEUDOSCORPIONES)

Few people have experienced the delight in seeing a pseudoscorpion, despite there being more than three thousand species worldwide, and most people don't even know they exist. That's partly due to their size, since most are no more than a couple millimeters long. They also rarely cross paths with humans, unless the human happens to be a naturalist looking at leaf litter.

While they are arachnids, pseudoscorpions are not scorpions, and they do not sting. They are tiny predators that prey on smaller organisms like springtails. They move a little like crabs, going backward and forward, pincers poised in the air. If you look closely at its tiny pincers, you may notice teeny hairs. Those hairs are sensitive to touch and help them overcome their poor vision, navigate their dark world, and find their springtail prey. Once they find prey, either by stalking or ambushing, they use poison glands in their pincers to subdue it and inject it with their saliva, which liquefies the insides, making them easier to eat.

Like some of their arachnid relatives, pseudoscorpions can produce silk. Unlike spiders, they spin the silk with their mouth and use it to construct small chambers with sand and plant matter. Here, they molt or hide from unfavorable weather. Because they're so small, they can't travel far, but they will hitch a ride with larger invertebrates, like passing flies, beetles, or harvestmen, by latching onto their legs—off they go for a new forest adventure.

PSEUDOCENTIPEDE (SYMPHYLA)

Here's another little-known invertebrate that resembles something more well known, although these tiny white invertebrates are only distantly related to centipedes and do not possess any venom. The symphylans, often called pseudocentipedes, are considered myriapods, the subphylum of centipedes and millipedes. They have twelve pairs of legs as adults but grow to only about 10 millimeters in length.

Symphylans are found deeper than leaf litter, in the top soil layers where they consume organic matter like decaying plants and fungi. Their role is that of decomposer, and they break down organic matter, releasing it back into the soil and enriching it. They are blind, relying on their beaded antennae to find their way around their compact world. Not large enough to create tunnels, they use their sinuous bodies and many legs to wiggle between gaps in the soil with great dexterity. When it is hot, they move deeper into the forest soil and have been found several feet down.

JUMPING BRISTLETAILS (MICROCORYPHIA)

Jumping bristletails spring about in the leaf litter and can leap several times their own body length, as far as 10 centimeters. They achieve this feat by contracting and then snapping their abdomen against the ground to launch up into the air. It's less elegant than jumping on legs perhaps, but highly effective.

Bristletails also run quickly, and there is evidence they can fly in a way, even without wings. Actually, they glide more like a flying squirrel, using long appendages that look like tails, to steer themselves. Interestingly, this ability may be a clue into how aerial insects gained the skill of flying.

Jumping bristletails can live up to a couple of years. They're related to silverfish, and in fact, the two were once grouped together but are now in separate orders.

Bristletails feed on lichens, algae, mosses, decaying plant matter, and fungi. Their eyes are large in relation to their head, which is small in relation to their body. Most of the roughly 450 species come in shades of brown and gray, fitting for hiding in leaf litter. But they don't limit themselves to forest floors. Bristletails have been found in the Arctic and in deserts.

Make a Pitfall Trap

SEASONS:
Spring, summer, autumn

STUDY TOPICS:
Ground invertebrate community, invertebrate observation

Hidden below the plants in the leaf litter is a liminal space, the place where soil meets the air. Plants and animals live here, but so do fungi and bacteria, not to mention protozoa. Plant matter decays even as new plant life rises up. Nutrients and rain enter the soil, fueling the growth, and that new life, in turn, fuels the invertebrate community. These few inches that cover the land across our planet contain a wealth of biodiversity. And yet this harmony does not just happen. It lives in a delicate balance that can, and often does, fall apart when disturbed by pesticides or clear-cuts.

Down at the surface, you'll find an entire host of organisms, from fungi to worms and beetles, working to decompose the dead. They process dead plants and animals—

A pitfall trap

without them, we'd be walking through carcasses instead of flowers. Alongside the decomposers are the earthmovers who circulate the precious, life-giving top few inches of soil. Worms and moles both do their part to rototill the soil, moving decomposed nutrients down and creating space for rain to filter in and reach plant roots. Meanwhile, predators like snakes and centipedes move around on the surface, feeding on other organisms and keeping their populations in check. Fungi, another predator, lives in the soil itself, a silent and deadly killer of many different invertebrates.

Despite the fact that we walk around on all this activity, we know little about it. Leonardo da Vinci once said, "We know more about the movement of celestial bodies than about the soil underfoot." And while we may understand more than we did in da Vinci's time, there is still a great deal to learn.

A naturalist can easily study and observe the lives of invertebrates in the soil using a pitfall trap, which is designed to capture creatures that wander around on the surface. The invertebrates fall in and become trapped. Pitfall traps can help you first learn which invertebrates are living in certain locations and, notably, which are absent. As we've seen, there are hundreds of human impacts on invertebrate populations, from pesticide use to climate change, and the invertebrate numbers, in general, are in steep decline. By documenting your findings, you can help provide important data on a vital part of the food web.

A pitfall trap can also show a naturalist what species are migrating. As our world warms, many species are on the move in search of cooler temperatures or food, or for other reasons we don't yet understand. Sharing the results of a pitfall trap with a community science project can make more data public and available for research. See the "Community Science Projects" section in Resources for project ideas and recommendations.

Setting up a pitfall trap is easy and uses materials you likely have on hand.

Materials

Container: The container needs to be deep enough that invertebrates can't easily crawl out of it. Repurposed food containers like yogurt cups, jars, cans, or other cuplike items will work.

Cover: You'll need something large enough to cover the cup to protect it from rain and falling debris, as well as hungry predators looking for an easy meal—a piece of wood, a tile, or anything that can act as a roof.

Rocks: A few small rocks relatively similar in size will help hold the cover in place but create space for invertebrates to move underneath.

Shovel: You'll need a small handheld shovel to dig a hole the size of your container.

Steps

Choose a Location: The best location is in soil hidden away under plants, where the invertebrates find shelter from predators. Avoid open areas like lawns where there is a lack of low-growing vegetation shelter or debris. Also avoid areas where pesticides are sprayed because most of those substances kill indiscriminately and you're unlikely to find many invertebrates.

Tip

If you have little to no success, you can try baiting the trap with a piece of ripe banana. To attract dung beetles, add pellets from rabbits or other herbivores. You can also try moving the trap to another location.

Dig: Use a shovel to dig a hole the depth and diameter of your chosen container.

Place the Container: When your hole is ready, place the container inside. Make sure it sits in so the top is even with the surrounding soil, not above or below it. Keeping it flush is very important. If it's too low, the soil will fall in, and if it's too high, the invertebrates won't fall in. Put a couple of leaves in the bottom for the invertebrates to hide under while they're trapped.

Place the Cover: Place the small rocks around the container and set your cover on top of the rocks. Leave enough room between the soil and cover for invertebrates to walk under, but not so much room that a hungry frog can easily enjoy a free buffet. Don't place so many rocks that the invertebrates can't get in.

Wait: Once it's set, let the trap sit overnight. Do not leave it longer, or the trapped invertebrates might die, or predators may discover them.

Check: Revisit the trap in the morning before the temperature gets too warm, which can overheat your invertebrates. Remove the cover and look in the container to see what has fallen in. You can use a magnifying glass or a hand lens. If you have a microscope, bring your container inside and place the invertebrates in a large petri dish with a lid.

Document: Make notes or take photos of what you catch so you can identify them later, and possibly contribute your records to community science. You can also sketch and make notes in your nature journal. Since the invertebrates are captured alive, this is a great opportunity to study their behavior.

> ## Tip
>
> Not only can you identify and document what you find while you're in the field, but you can also bring some of what you find inside for further observation. See Project 12, about making a habitat terrarium, for more long-term observation.

You can put each in a separate petri dish with a leaf and watch them explore their surroundings. How do they walk and move? Study their anatomy and the way their legs bend, how their antennae move, and how flexible, or not, their bodies are. Do they scurry around looking for somewhere to hide, or do they freeze and refuse to move?

Identify: There are many useful guidebooks about backyard invertebrates, including region-specific guides, which will help narrow down your choices when trying to identify your local invertebrates. An excellent resource is BugGuide, a website managed by entomologists who help identify requests submitted.

Release: When you're finished observing your invertebrates, release them back outside where you found them.

Experiment: If you set your pitfall trap in the same location over a period of time, you can mark the invertebrates that fall into the trap with a dot of nontoxic, waterproof paint to see if the same individuals fall in again.

Tip

Try setting out your pitfall trap regularly throughout the year to see which invertebrates show up during the different seasons, and try a variety of locations.

Ground Invertebrates

You might encounter a wide variety of invertebrates in your pitfall trap, not all of which will be insects. This project will capture the invertebrates that cannot fly or jump to escape, which may include spiders, mites and harvestmen, centipedes and millipedes, silverfish, ants, worms and springtails, and a few special species described below.

WOODLOUSE (ONISCIDEA)

Often called the roly-poly or pill bug for their common defense of rolling up into a ball to defend against predators, the woodlouse is one of the most familiar and beloved of our backyard invertebrates. Woodlice are not insects, however; they are isopods, which are crustaceans, and more closely related to crabs and shrimp than other invertebrates in most peoples' backyards.

Woodlice are found only where it is moist because they breathe through special appendages that diffuse gas

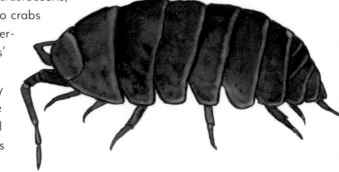

through water, a remnant of their ancestral life in the sea. The copper in their blood is another remnant of their time underwater. They need it to absorb oxygen, but every time they defecate, they lose a little bit. So woodlice have developed a curious habit of eating their own dung (known as coprophagia), which helps them restore their copper levels. But that's not all they consume.

Woodlice are detritus feeders, meaning they eat decaying material but will also eat lichens and algae. We can thank them for breaking down newly fallen leaves and starting the process of decay for the invertebrates known as secondary decomposers. Springtails, mites, and bacteria move in after woodlice are done to finish the process of returning leaves to the soil.

Despite the roll-up-into-a-ball defense (which not all species can do), other invertebrates still prey on them, including centipedes, beetles, and spiders.

EARWIG (DERMAPTERA)

The first thing most people notice about an earwig is the long, sicklelike pincers at the end of their elongated, slender body. But look closer and you'll find that earwigs are beautifully colored in shades of mahogany and maroon. They are sleek and shiny, with hidden wings folded like origami under a pair of armor-like forewings. In the rare instances when they stretch out them out, their hidden wings unfold dramatically, to be up to ten times the size they are when folded.

Unlike the vast majority of insects, earwig mothers actively care for their offspring. The mothers first lick their eggs to prevent mold from growing on them, and once the eggs hatch, they tend to the nymphs in the nest until the young ones have molted a couple of times.

Earwigs are omnivores. Like woodlice, they are decomposers and help break down plant debris, but they won't pass by an opportunity to consume fungi. A few are even predators, searching for small invertebrates. They're active mostly at night, spending their days cleverly hiding.

Earwigs are thigmotactic, meaning they like to be in close contact with surfaces, top and bottom. This is why they hide under plant containers and leaves, or squeeze into other tight spaces on the ground, preferably somewhere moist. This curious behavior helps them stay safe from their many predators. Birds, spiders, beetles, and other invertebrates would all happily consume an earwig.

If they're caught out in the open, though, the earwig is not defenseless. They will first use their cerci, the pincerlike appendages on their backside, to deliver a strong nip. Failing that, the earwig will squirt out a foul-smelling fluid to deter any would-be predator. You can tell a male from a female because the female's cerci are straighter and closer together, while a male's are more curved.

NIGHT-STALKING TIGER BEETLE (OMUS SPP.)

A creature of the night, this tiger beetle has a fitting appearance. At first, it looks large and black. However, cast a little light on the roughly textured body and you'll discover a subtle yet beautiful purple, iridescent sheen. These mostly nocturnal beetles can sometimes be active during the day, but they're more likely hiding under leaf litter or logs, just one of many ground beetles who hide during the day and come out to hunt at night. Others include such incredible names as the long-faced snail-eating beetles, gazelle beetles, carrion beetles, stag beetles, and rove beetles.

Many ground beetles are carnivorous. The snail-eating beetles have elongated heads shaped to reach inside snail shells. Others are detritivores, which consume animal dung and other decaying material. Some, like rove beetles, are omnivorous, feeding on other invertebrates, fungi, decaying material, and even carrion. Night-stalking beetles feed on small invertebrates.

Night-stalking beetles create burrows underground where their larvae live for as long as three years. The larvae wait for prey to pass by the entrance of their burrow. When an ant, for example, happens by, the fierce larvae grabs it and drags it down into the burrow to eat it. As adults, the beetles use their large, pinching mandibles to attack anything they can grab. But they don't rely only on their mandibles. Some night beetles excrete an enzyme that breaks down the body of their prey's exoskeleton. Once softened, the tiger beetle can easily chew it up. The enzyme can also be used as a chemical defense from their own predators, which include venomous robber flies, dragonflies, lizards, salamanders, and even small mammals.

WOLF SPIDER (LYCOSIDAE)

These fast spiders spend most of their time on the ground and don't spin webs. Instead, they chase down their prey and pounce to catch them, much like a wolf. They'll eat most anything they can, including ants, grasshoppers, and flies. Wolf spiders are found in most habitats and are often dark and mottled brown or gray to blend in with the ground, making them difficult to see unless they are moving.

Doting mothers, the females spin a sac around their eggs after mating and carry it under their abdomens. After the spiderlings hatch, they ride around on their mother's abdomen. Wolf spiders are the only spiders known to care for their offspring in this way.

Like most spiders, they are very shy and spend most of their time, especially during the day, hiding. Wolf spiders have excellent short-range vision for detecting moving prey and can be identified by their unique eye arrangement. Four small eyes line the lower part of their face, while a pair of large eyes sits over them. They have another large pair on top of their head behind their front-facing eyes. Their excellent vision doesn't always prevent them from becoming prey themselves, and wasps often capture wolf spiders to feed to their larvae. In addition to wasps, birds, reptiles, amphibians, and small mammals may feed on wolf spiders.

PROJECT 11

Go Bush Beating

Shrubs exist in a middle ground—less majestic than upper-story trees, and less sought after than the herbaceous flowers at ground level. These middle children of the plant world deserve respect, however, as a vital part of the ecosystem. Shrubs combine the best qualities of the other two layers by providing habitat with woody structures and producing beautiful flowers, the combination of which offers shelter and food for wildlife.

Those aren't their only services either. Bushes are often one of the first plants of any size to appear in ecological succession, which is how a species community changes over time in a habitat, and they continue to influence what wildlife is found in a particular ecosystem. They reduce air flow, which decreases water loss from surrounding

123

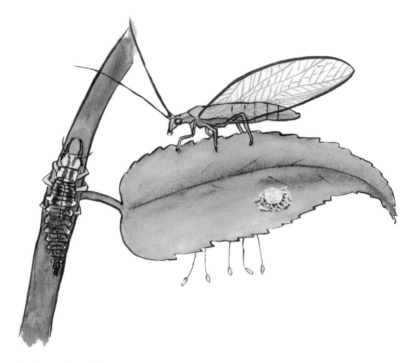

Lifecycle of a green lacewing

plants, and they provide shade. When a bush's leaves fall, they decompose, adding important organic matter and nutrients to the soil, while evergreen shrubs offer shelter for wildlife all year long. In some habitats, shrubs are the tallest plants that grow, and they fill a critical role in the ecosystem as the only shelter and food for wildlife and nesting place for birds. Shrubs are a wildlife buffet, in fact—food for everything from the tiniest invertebrates to the largest bears. Flowers are pollinated by insects and often turn into fruit that birds and other wildlife consume.

Each individual shrub is a microhabitat containing an entire world. Many invertebrates have some type of relationship with shrubs. Some, like the tardigrades we saw living in the moss and lichen in Project 6, use them as a place to hide, while others use them for food. In fact, invertebrates that live in shrubs eat most parts of the plant, from root to leaf, before becoming a vital source of the food web themselves. Migratory birds visit shrubs to feed on the bounty of herbivorous invertebrates, and parasites are silently waiting. And then there are the invertebrates who don't feed on the plant but instead eat each other. Life for invertebrates in a shrub microhabitat is perilous, and death rarely comes from old age.

It's easy to see butterflies, bees, and flies visiting a shrub's flowers, but invertebrates living unseen among the roots and branches are far less obvious. A naturalist is more likely to discover some of the hundreds of signs that invertebrates leave behind. Holes in leaves may indicate caterpillars, sawfly larvae, beetles, slugs, or leafcutting bees. Tunnels in leaves can reveal the presence of leaf miners like the larvae of moths, flies, or beetles. Strange spots may indicate leafhoppers or aphids, and rolled leaves may be made by caterpillars, spiders, or other invertebrates. A naturalist may also encounter wildly diverse invertebrate eggs, many of which are laid directly on plants.

Invertebrates compose 97 percent of all life on Earth. With that massive diversity, they live absolutely everywhere, and naturalists need an array of techniques to find and learn about them. There are two types of methods for searching for invertebrates: passive and active. Passive methods include finding leaf litter and soil invertebrates with the Berlese funnel trap (see Project 9) and ground invertebrates with the pitfall trap (see Project 10). Then there are active methods like bush beating. When it comes to discovering the microbeasts hidden away in vegetation, you can't beat around the bush—you have to beat the bush. Literally.

Bush beating, also known as tray beating, is an active method that naturalists and scientists use to find invertebrates on plants. It's a simple, common, and standard technique that involves hitting a plant with a stick so the invertebrates fall onto a prepared tray.

Knocking invertebrates out of a shrub will give you an excellent idea of its inhabitants while revealing insight into the plant's natural history. Bush beating is easy and requires only a few materials. You can do it anytime of day throughout the entire year, although the results won't be ideal if it's raining or windy or if the plant is wet.

Materials

Tray, Sheet, Paper, or Umbrella: There are many options for catching the invertebrates. Scientists often use specialized fabric trays, which look like a square of white fabric with pockets in the corners to tuck in the ends of tubes or dowels, making an X, very much like a kite.

Make an Aspirator

You'll need a clear container with a lid, a piece of 10-millimeter rubber tubing, a small piece of fabric, and a rubber band. Cut the rubber tubing into two pieces. The tube you suck on only needs to be a couple of inches long, but the tube to suck up the invertebrate should be a bit longer, about two feet, so you can easily reach them.

Cut or poke two holes the size of the tube diameters in the container lid. Push each tube about a half inch through the holes in the lid and glue them in place. Put a small piece of fabric over the end of the shorter tube, on the inside of the lid, and attach it with a rubber band. When you inhale, the fabric will prevent you from swallowing the invertebrate. Put the lid back on the container and you're ready to go.

You can also use a piece of white fabric, such as a sheet, laid on the ground under the shrub's branches. For very small beatings, such as a single flower or seed head, a piece of white paper works well. Another clever option is to turn an umbrella upside down under the shrub. It even folds up small and is easy to carry and use in the field. Regardless of what you choose, make sure it's white or very light colored so you can spot the invertebrates easily.

Stick: Any sturdy stick at least a foot long will work. Don't choose something so large it will break the branches or otherwise damage the plant.

Containers: If you want to observe or document any of the invertebrates (especially predatory invertebrates) you knock out of the shrub, you'll need individual containers to put them into before they escape or eat one another. Any sort of jar or clear, lidded container will work for short observation.

Aspirator (optional): If you want to collect the invertebrates you knock down, an aspirator, also called a pooter, could help. An aspirator is a simple device with two tubes attached to a container. You inhale on the end of one tube and hold the other over the invertebrate, which acts as a vacuum, sucking the invertebrate into the container. This is a good tool for small invertebrates that may get damaged if handled with hands or forceps. Be careful not to suck too hard on the smallest, most delicate invertebrates, though, because the pressure can crush them. You can buy aspirators from entomological supply stores or easily make your own (see the bonus activity).

> ## Warning
>
> Know your plants, and avoid touching any that are poisonous, like poison oak, or any that have thorns, spines, or other features that may cause injury.

Steps

Choose a Shrub: While invertebrates can be found virtually everywhere, some shrubs host more than others. Identify a shrub of choice, and reference plant books specific to your region to learn more about it if you're not already familiar. Find out what the flowers look like, how large it grows, what wildlife it attracts, what habitat it's usually found in, and so on.

Look for plants native to your region because they will have more diverse invertebrates than cultivated and introduced plants. Avoid areas where pesticides are sprayed or applied.

Place the Tray: Depending on what type of tray you're using, either hold it or lay it down under the plant.

> ## Tip
>
> This method isn't limited to shrubs. You can also try it on low-hanging tree branches or with grasses and flower heads, albeit with a more delicate touch. It's also possible to use this technique on dead wood.

Hit: Give the branch a sharp rap with your stick, hard enough to shake it, but not hard enough to break or damage the plant. If you hit too gently, you'll mimic the wind, and the invertebrates will hold on tighter.

Study: Look fast, because the more mobile and flying invertebrates will likely jump or fly right out of your tray. If you want to observe the invertebrates or document the ones you found, collect them into containers quickly so you can look at them each in turn.

Once they are collected, study them to try to identify what invertebrate families are represented. This is a good time to reference your regional field guides with the invertebrates in hand. Studying these guides ahead of time will help you have an idea of what the different families look like. If you're stumped or want to try and get closer to the species level, websites like BugGuide can help with identification.

Observe each invertebrate and note their anatomy. How many legs do they have? That's not always such a simple question to answer since some, like caterpillar larvae, have prolegs, fleshy, stubby legs farther down the body that don't count toward the standard six legs all insects have. What size are their eyes, and how many do they have? Then study their behavior. Are they trying to hide, or are they running in circles, trying to find a way out?

Perhaps you'll see something you don't recognize, like a larvae cocooned in a white web, strange silky protrusions on a comatose caterpillar, or an ant with a bizarre stalk emanating from its head—all signs that parasites, whether invertebrate or fungal, are also living in the shrub.

Document what you find, and compare your findings over time and across locations.

Warning

Depending on where you live, exercise caution when handling invertebrates with bare hands, and don't touch anything unfamiliar. Some invertebrates may bite or sting when handled, and others, like caterpillars, may have stinging hairs.

Return the Invertebrates: Once you're done observing and documenting the invertebrates, shake them back into the shrub they came from so they can continue on with their brief lives.

Repeat: Try this method in different habitats, from backyard to forest and everywhere in between. Also try it at different times of year, even in winter, when different invertebrates will be active. Survey not only different plants but also the same plant throughout the year to see how the invertebrate population changes through the seasons. Try different species of shrubs because many invertebrates, especially sawfly and butterfly and moth larvae, have specific host plants, the genus or species they prefer to feed on.

Shrub Invertebrates

A naturalist may knock any number of countless invertebrates out of a bush. Some might be there briefly, simply resting or hiding from hungry birds. Others, like sawflies, butterflies, or lacewings, might be looking for a place to lay eggs, or their larvae may be busy chewing on leaves. Insects such as aphids and leafhoppers might be sucking fluids out of plants. Parasites like wasps may be looking for an invertebrate host for their own offspring, and predators like lady beetles and spiders might be looking for herbivore invertebrates to consume. You may even discover cocoons of moths, flies, or other invertebrates. Keep an eye out for parasitized invertebrates as well.

Here are a few of the more common invertebrates you may encounter.

LEAF BEETLE (CHRYSOMELIDAE)

With an estimated sixty thousand species of leaf beetles in the world, you'll likely encounter at least one at some point when bush beating. The beetles of the Chrysomelidae family are extremely common on plants because they are phytophagous, meaning they feed on living plant material. Adults consume pollen and leaves, while their larvae tend to live beneath the surface, feeding on plant roots. Some are leaf miners, living inside a leaf. Leaf beetle species tend to be host-specific and will be found only on certain plants or a few species of plants.

Among beetles, leaf beetles are small, and their distinguishing identification features are difficult to see. Their oval or elongated bodies are usually no longer than 12 millimeters. Some stand out with their bright colors or metallic sheen, but many are more subdued shades of black, brown, or green. Some of the tortoise beetles are very striking with clear-edged shells. Because of the diversity in their appearance, leaf beetles are easily confused for other beetles, like lady beetles and long-horned beetles.

Leaf beetles cleverly take in chemicals from the plants they eat and use them for their own defenses by releasing a toxic secretion as a repellent. One species of tortoise beetle takes this a step further, collecting their own feces and carrying the toxic waste on their backs as a defense. That's a shield that no superhero can top.

SAWFLY (SYMPHYTA)

An invertebrate that almost seems to be designed to confuse naturalists, the sawfly adult looks like a mix between a bee and a fly. These curious invertebrates may resemble and share a name with flies, but they're actually in the order Hymenoptera and more closely related to wasps, bees, and ants. Like bees, sawfly adults have four wings and long antennae, but they have a thick waist more like flies, and they do not sting.

You're more likely to knock the larvae of sawflies, which look like butterfly or moth caterpillars, out of a shrub than the adults. Sawfly adults feed on pollen and nectar but are not active like bees so it's harder to encounter them by chance, but like their larvae, they can be found reliably around their host plant.

Sawflies live as adults for only a matter of days. Most adults are female because they are parthenogenic, meaning they don't need males to reproduce. The female's ovipositor is sawlike, giving them their name. They use this apparatus to cut a slit in the leaves, new shoots, or stems of plants to lay their eggs.

Unlike both flies and bees, sawfly larvae feed almost exclusively on leaves or conifer needles, and some bore into stems. There are multiple families among sawflies, and almost all can be found on vegetation, with the exception of one parasitic group. Many species cluster together as young larvae and disperse as they grow. Sawfly larvae and butterfly and moth caterpillars all have three pairs of true legs closest to their head, but sawfly larvae have at least six pairs of prolegs, while butterfly and moth larvae have no more than two to five pairs. Sawfly larvae have very little to no hair. Unlike caterpillars, sawfly larvae readily rear back, lifting their abdomen, as a defensive posture.

LACEWING (NEUROPTERA)

The adult lacewing is as dainty as the larvae is fierce. It's hard to believe they're the same insect. Lacewing larvae are highly efficient predators that resemble little alligators. They feed on insects, many of which humans usually consider garden pests, like aphids and mealybugs, as well as caterpillars and whatever else they can catch with their long, scythe-like pincers. Some green lacewing larvae carry the corpses of their prey on their backs for camouflage, like grotesque trophies. Other green lacewing larvae 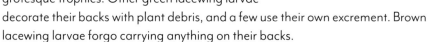 decorate their backs with plant debris, and a few use their own excrement. Brown lacewing larvae forgo carrying anything on their backs.

Green lacewing adults are not strictly predators like their larvae and may be herbivores or opportunistic omnivores. Herbivorous adults generally feed on nectar, pollen, or the honeydew produced by aphids. Brown lacewing adults are predators. While lacewing larvae are active during the day, adults tend to be crepuscular, or nocturnal, so they aren't readily seen. The most commonly encountered types are green lacewings (Chrysopidae family) and the smaller brown lacewings (Hemerobiidae family), but there are a few other, more obscure families of lacewings as well. Green lacewings lay a single egg at the end of a very long, slender stalk that is attached to a plant, while brown lacewings attach theirs directly to leaves without the stalk.

WEEVIL (CURCULIONOIDEA)

See no weevil? Not likely. The weevil family Curculionidae is arguably the largest in the animal kingdom, and with more than fifty thousand species in the world, chances are good that your bush beating will turn up a weevil.

Weevils are defined by their impressive snouts, called rostrums, which are often long and slender with a mouth way out on the end. Some weevils do not possess the iconic snout, such as the bark beetle, but the majority do. Because most weevils are sexually dimorphic (meaning females and males are different in appearance), you might also notice that females tend to have longer rostrums than males. They are not a particularly colorful group of beetles, and the vast majority are fairly small and come in shades of brown or black. Larvae are largely nondescript grubs of a beige color with a brown head.

Nearly all weevils are herbivores and live their entire lives on or in plants. Virtually no part of any plant is immune to weevils, and adults or larvae can be found in or on the stems, flower buds, leaves, roots, fruits, nuts, and just about everywhere else. Most weevils have a host plant, and those plants range from trees to grasses and everything in between, including some aquatic plants. A few species bore into wood and feed on the fungi that grows in the tunnels they excavate. Their resourcefulness may help explain their massive diversity.

Weevils are the opossum of the invertebrate world; when they are disturbed, they play dead, remaining motionless on their backs until danger has passed.

Make a Habitat Terrarium

SEASONS:
All

STUDY TOPICS:
Invertebrate life cycle,
plant growth

Sometime in 1829 a naturalist encountered the pupa of a sphinx moth and wanted to see the adult that would emerge from it. He found a large glass bottle, put some soil in the bottom, and buried the pupa, mimicking the way that some species of moth pupae often hide in soil. Then he waited. Soon he noticed something beginning to emerge from the soil. It wasn't a moth, but plant sprouts. With true naturalist fascination, he turned his attention to the plants growing in the bottle and began to study them. The moth never emerged, but grasses and ferns did, along with a new lifetime's worth of work.

The naturalist was Nathaniel Bagshaw Ward, a physician from London and the man who brought the first terrarium to the world. His accidental invention sat in his

A habitat terrarium

windowsill for several years while he dutifully documented the growth and development of grasses and ferns and then wrote a book and created a tool that would usher in a new era of transporting live plants around the world.

His invention, the Wardian case, had a wooden base with glass panels, and was specially designed to protect live plants during the era of monthslong ship transportation. During a time when botanists were lucky if one or two plants survived a long voyage, Ward's invention reversed those statistics. Inside the case, very few plants would die. The case changed the botanical world, and evolved into the ornate and beautiful glass and metal terrariums for ferns and other plants that were found in many Victorian homes.

These days, actual Wardian cases are rather obscure, but they have expanded into countless types of its offshoot, the terrarium. In the basic sense, a terrarium is any enclosed space used to grow plants, especially those that need special conditions that cannot be achieved in a simple container.

Where Ward failed in rearing a moth, another naturalist not only succeeded but also mastered the insect version of the Wardian case. Just a few years after Ward published *On the Growth of Plants in Closely Glazed Cases*, Henry Noel Humphreys published his own treatise titled *The Butterfly Vivarium; or, Insect Home: Being an Account of a New Method of Observing the Curious Metamorphoses of Some of the Most Beautiful of Our Native Insects*. Humphreys was an incredibly accomplished artist and naturalist with a fondness for insects, particularly moths and butterflies. He considered the established rearing cages "rude," nothing more than functional wooden boxes. With his artistic eye, he enhanced Ward's idea by redesigning them into elegant cases for raising Lepidoptera, the order of insects that includes butterflies and moths, and observing the wondrous transformation of a caterpillar into a winged creature. He went on to make a case with room for rearing aquatic insects as well as terrestrial ones by adding a water reservoir. Today, it's possible to make a specialized terrarium to raise invertebrates and small vertebrates like snakes and frogs, called a vivarium.

The benefits of a terrarium for a naturalist are many. It is the perfect tool for naturalists to grow a plant or replicate a habitat for close observation. We can study habitats that may not exist where we live, like deserts or tropical forests, right in our own home. But we can also use a terrarium to study our local habitats in greater depth. We can watch the plants develop from seedlings to maturity or grow lichens, mosses, and other cryptogams. We can put garden or forest soil in a container like Ward did and see what emerges. Put some dead wood in it, and slime molds or fungi may sprout up.

Terrariums can also serve as the terrestrial version of the wetland habitat featured in Project 3. Once you build a habitat terrarium, you can temporarily place invertebrates that you encounter and want to observe more closely, like woodlice, beetles, spiders, or crickets. Want to watch a slug eat? Put it in the terrarium with some food and observe. Want to see how a cricket sings? An enclosed terrarium makes an ideal observation chamber.

Terrariums can also be fine-tuned to keep local invertebrates for longer periods of study. While observing out in nature is ideal, replicating that habitat and conditions can make observation possible indoors, year-round, and offer a view of behavior that might be hard to observe in the wild, like feeding, web making, mating, and even metamorphosis. It brings plants and animals closer and allows the naturalist to not only observe their behavior but also understand their life cycles. An enclosed habitat also provides a close, detailed look for sketching purposes, allowing a naturalist to study invertebrate anatomy, behavior, and more at their leisure.

Materials

You can make your own habitat terrarium to observe local organisms with a few materials and a little creativity. There is a lot of flex-ibility and different options, depending on your goals and whether you want something simple and temporary for a single project or something that will last long term.

Note

When collecting plants, always be sure to collect ethically and legally, follow-ing local regulations.

Container: When it comes to choosing a container, the only rule is that it should be clear enough to see through. An arid terrarium doesn't need a lid, but if your habitat requires humidity, or if you're keeping invertebrates inside, you'll need a lid of some type. Avoid containers with colored glass, which can impact the intensity of the light on the plants and possibly block or increase certain parts of the light spectrum.

You could buy a container specifically designed as a terrarium, but a mason jar or soda bottle can work perfectly well, or you could find some other nonconven-tional type of suitable container at your local thrift store. If your goal is to observe small invertebrates, don't choose a large container where they will get easily lost, and consider what shape will work for the particular type of invertebrate. For exam-ple, spiders, flies, and other aerial invertebrates need a container that's taller than it is wide so they have room to move, while soil- and ground-dwelling invertebrates need a container that's wider than it is tall.

You can poke a few tiny holes in the container, but humidity is more import-ant than ventilation. Most invertebrates absorb moisture through their bodies and don't actively drink. Too much ventilation will allow moisture to escape, and the terrarium will dry out. Invertebrates won't suffocate as long as long as the container isn't airtight.

Gravel: Gravel, stones, or other small rocks allow excess water to drain away from the plants. The rocks should be small enough that the soil does not fall down into them. Aquarium gravel works well, or pea gravel depending on the size of your terrarium, but sand is too small.

Tip

Avoid containers built to be decorative and not functional. They may not be waterproof, and moisture could leak and ruin the surface it is sitting on.

Horticultural Charcoal (optional): A layer of charcoal between the gravel and soil helps prevent mold and algae from growing in the damp terrarium. Bags can be found in gardening stores.

Soil: The type of soil depends on what you want to grow. Potting soil is useful for most applications, but many terrarium instructions recommend a mix of soil, sand, perlite, and other components. Garden soil may contain unexpected bacteria, invertebrates, and plants, which to a naturalist is a huge bonus, so feel free to start with garden or forest soil from the habitat type you're trying to replicate. If you only have access to backyard garden soil, that's fine too. You may end up with nematodes, ants, woodlice, earthworms, springtails, or larvae or pupae of different invertebrates.

Warning

Peat moss is commonly used in terrariums and other gardening projects, but peat bog habitats are disappearing at a devastating rate because of excessive harvesting. Peat bogs are incredibly slow to grow and also happen to be the largest terrestrial storage of carbon, helping to mitigate climate change. Damaging them releases massive amounts of greenhouse gases. To help preserve this important, irreplaceable habitat, please avoid using peat moss.

Plants: Choose suitable plants based on the type of terrarium habitat. If you're making a temperate woodland habitat, select ferns, mosses, and even lichens. For a desert habitat, succulents and small cacti are great options. The key is not to mix and match plants

> **Note**
>
> Most invertebrates don't drink water; they absorb it from the moisture around them. But there's no reason you can't add a small water element if you want to, or even make a complex half-terrestrial, half-aquatic system.

with different requirements. If you're making a local habitat terrarium, collect small plants from the area and experiment to find what works well. Also be sure to choose plants that stay small; however, you can also prune some plants regularly to fit.

Habitat Elements: A habitat is more than just the soil and plants; it may include dead bark, branches, or other bits of wood, rocks, dead leaves, cones, seed pods, and even bones and empty shells. The dead material is important for providing shelter, perching locations, and food for invertebrates and fertilizer for the plants.

Distilled Water (optional): Tap water can leave white residue on your terrarium when you water it. You can use distilled water to avoid buildup on the glass.

Steps

Because there is no one-size-fits-all approach to a terrarium project, you'll need to do a little preliminary research to make a suitable home for local invertebrates. The first step is to observe and document the details of the habitat where you find them. A spider in a web strung across two high branches may not like an enclosed, humid terrarium, but an airy jar with sticks in it may suit it very well. The woodlouse found hidden under a piece of bark in damp moss will not like an open, dry terrarium but may thrive in a closed, temperate terrarium.

Clean Materials: Mold and algae can ruin a terrarium quickly. Thoroughly clean your container before assembling it.

Add Base Materials: Lay a couple inches of gravel at the bottom followed by a half-inch layer of charcoal. Place the soil over the charcoal. It can help to moisten the soil so it's not dusty when you place it in, but don't make it so wet that you can squeeze water out of it.

Add Plants: Remove each plant from the soil it was in. Gently untangle the roots, and then plant it into the soil in the terrarium.

Add Water: Because tap water can leave residue on glass, use distilled water to water the terrarium. After all the plants are installed, spray the soil thoroughly with water. It should be moist but not saturated.

Place: Do not place your terrarium in direct sunlight, which will cause it to get too hot. However, it will need regular, indirect light. A lamp can provide suitable light if you do not have a good location in range of natural light.

Add Invertebrates: Once the terrarium is established, begin to add the invertebrates you wish to observe and study. You can put them inside it for just a few hours and then return them to the wild, or you can maintain the habitat to watch several generations. Be sure to match the invertebrates to the habitat type, and as much as possible, do not unduly stress them. If you add multiple species, be mindful of predators that may kill your other invertebrates. The idea is to gently observe—not create gladiator death matches.

Feed Invertebrates: If you're including invertebrates, be sure to find out their food source and provide it for them. Some invertebrates need fresh plants, others need decomposing materials, and predators need their prey source.

Maintain: A terrarium needs little care once it's established. A closed terrarium generally needs water every few months, when there is no more condensation on the container or plants begin to wilt. Don't allow it to dry out completely. Mist it with a sprayer of distilled water instead of pouring it in. It's virtually impossible to get rid of excess water, so always err on the side of too little.

Warning

If you add invertebrates to your terrarium, even temporarily, be sure those species aren't endangered, at risk, or otherwise threatened. Most invertebrates aren't protected by laws, but naturalists should always practice sound ethics when studying and collecting native plants and animals. If you can't be fairly confident that you can keep it alive, leave it in the wild until you have done enough research and observations to care for it.

Preserve a Spiderweb

SEASON:

Autumn

STUDY TOPICS:

Spiderweb anatomy,
silk composition

F our hundred million years is a long time to master something, so it's no wonder that the spiderweb is the epitome of animal structures. Although a few other invertebrates use silk, none have perfected the material quite like spiders, who are artist, architect, and engineer all in one. Their diverse webs are as effective and deadly as any mammalian predator— perhaps more so, in fact.

Dew-coated webs that stretch across branches on foggy mornings are a particularly beautiful autumnal sight for us, but for a spider, the web is a life-dependent structure. Without a functional web, spiders would starve before they accomplish their most important goal, finding a mate so that they can pass on their genes.

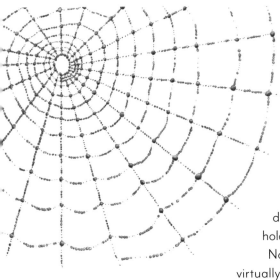

A dew-covered orb web

As many as fifty thousand species of spiders spin up to 130 different types of webs. The vast majority of humans are most familiar with the classic orb web, but spider architects have come up with many designs that may catch the attention of naturalists around the world. Some webs are densely woven funnels; others look like triangles, tubes, or three-dimensional mesh; and some spiders hold their web in their legs to cast onto prey. Not all spider species build webs, but virtually all of them use silk in one way or another. Some use it to actively hunt, others create draglines to catch themselves if they fall, and many use it to house and protect their precious eggs. Some spiders make a protective platform where they can molt safely, and others build temporary little nests that act like sleeping bags where they hide during the day. The diving bell spider constructs a structure of silk to collect air that it uses to breathe underwater.

One ingenious way that spiders use silk is to fly. In a technique called ballooning, they climb up a plant or other surface, point their abdomen into the air, and release silk to let the wind catch it and carry them up and into the sky. There is some evidence they make use of Earth's electrical field as well as the wind. Sometimes spiders "fly" in masses, with thousands taking to the air at the same time—and they can travel far. Spiders have been found up to two miles up in the sky, which helps explain how spiders are among the first species to cross seas and arrive on newly formed volcanic islands.

Some spiders also use silk as a way to communicate, like a scented billboard. Web-spinning female spiders weave pheromones into their webs for potential mates to detect. Different species have different pheromones, which males can differentiate, while the web itself serves as a sort of courtship location, like a restaurant for a blind date. The male and female meet to see if they're compatible. The male may dance, plucking at her silk threads with his feet. Or he may bring a gift of some food. If they hit it off, the male may add his own silken touches to her web before he makes his exit. But if he doesn't meet her standards, she may just eat him.

Spiders, as a group, have seven spinnerets, their silk-spinning organ, each associated with a different type of silk gland. No spider has them all, though some have up to five. They use these different types of silk for different functions. Some glands produce silk used to spin the sticky part of a web; other glands produce silk used to wrap up eggs.

A miraculous material, silk is a liquid until a spider uses its legs to pull it out of its spinnerets. No other natural fibers compares with the elasticity, strength, and toughness of spider silk, which can be five times stronger than steel and twice as flexible as nylon. It can also be stretched incredibly far and then released while staying taut. The elasticity allows the silk to dissipate the energy of the impact from an insect flying directly into it, without breaking.

Looking at a spiderweb under a microscope, you will probably observe a thin thread lined with clear droplets. These sticky globules capture and hold the prey that blunders into the web. Another type of spiderweb silk is not sticky at all, but instead has a dry, woolly texture almost like yarn. Insects still get stuck in this cribellate silk, a mass of extremely thin threads called nanofibers, hundreds of times thinner than regular web silk. It helps ensnare insects by sucking waxy chemicals from the prey's exoskeleton and infusing them into the web, helping to create a sort of rope. Essentially, the silk uses the prey's own body to trap it.

Spider silk has long been of interest to scientists and naturalists, who are continually learning new things about this wondrous material. A naturalist at home can easily observe spider silk and the construction of webs by capturing them on paper or studying a few strands under a microscope.

Materials

Paper: Cardstock, cardboard, or other heavy-duty paper will work best for capturing a web. Choose a color that contrasts with the color spray paint you'll use, and feel free to get creative. You'll also need old packing paper or newspaper to keep from getting spray paint everywhere.

Spray Paint: Look for a nontoxic spray paint in a color of your choice. White or black offer the best contrast for viewing on paper, but feel free to experiment and be creative.

Scissors: It's possible to sever the silk thread by simply pushing against it, but for a clean, intact web, you'll want scissors to snip through the silk strands. It's always handy to keep a pair of compact scissors in your field bag, like those found on a pocketknife.

Artist's Fixative Spray: To keep the web in place on the paper and preserve it, you'll need an artist's fixative, found in any art supply store. Hair spray can work as well.

Microscope, Slides, and Clear Nail Polish (optional): If you want to examine different webs and types of silk closely, you will need a microscope and basic slides with cover slips. A compound microscope works best to study the silk. You can use clear nail polish to mount the web to the slide.

Steps

Search for Webs: In the Northern Hemisphere, autumn is synonymous with spiderwebs for good reason. It's the time of year spiders are fully grown and making large webs. You can find spiderwebs in any season, but autumn offers the best abundance. Of course, autumn also brings rain, and wet webs are harder to work with. Look for dry webs around porch lights, on trees, in shrubs, and even low down on grass.

Once you find a high-quality web, make sure that it's uninhabited. A spider with a web will always be touching it in some way. Look for its creator along the edges where it's tethered to trees or branches. It takes some species several days to spin a new web, so you don't want to destroy an active web and disrupt their lives.

Prepare Paper and Web: Hold a large piece of packing paper or newspaper behind the web, then spray it with the paint color of your choice. Repeat on the other side of the web.

Capture: There are two approaches to capturing the web, depending on how much time you have. For the first, quicker approach, after spraying the web with paint, carefully press the paper against it before the paint can dry. Once the web is on the paper, snip the anchor threads so it comes free without disturbing it.

The second approach is to paint the web and leave it to dry for an hour or so. This way there's no risk of the paint smearing. Once it's dry, spray the web with fixative or hair spray to make it sticky again, and press the paper against it in the same way.

After capturing the web, allow it to set fully and dry.

Preserve: Once you've successfully captured it, spray a protective layer of artist's fixative over the web and let it air-dry in a well-ventilated space. Then make notes on the paper about where and when you captured the web, the spider type if you know it, the type of habitat, and any other relevant data for your records.

Collect on a Microscope Slide (optional): Paint a thin layer of clear nail polish on the middle of a slide, about the size of a cover slip, and allow it to air-dry for a minute. Avoid touching the polish. Carefully press the polished side against an interesting part of the web. Remove any excess web and place a cover slip on top to seal it in. Look at your specimen under the microscope at different magnifications to observe the construction of a single silk strand.

Tip

If you don't want to collect a spiderweb but simply want to observe it in nature, there are techniques to make a web easier to see in the field. One method is to mist it with water from a small spray bottle. Another is to use cornstarch to dust the web. You can carry cornstarch in a sock and tap it out over a web. If you carry the sock in your field bag, put it in a plastic bag so it doesn't make a mess.

Another mess-free option is to repurpose a baby powder container as a cornstarch carrier. Don't use baby powder itself, though, because it can harm spiders and other living creatures in the area.

Common Spiderwebs

The classic web of the orb weaver spiders is merely one design. Spiders can create webs that look like tangled piles with seemingly little rhyme or reason, flat sheets resembling hammocks, or elaborate funnels seemingly spun in a fairy tale. Not all make suitable candidates for preserving, but all are worth recognizing and studying. Here are a few you may encounter.

ORB WEB (ARANEIDAE)

Undoubtedly the most iconic of
all webs, these circular structures are
made by spiders in the Araneidae family,
appropriately known as the orb weavers.
There are about thirty-five hundred species
in this family around the world, with bodies
ranging in size from a quarter inch to well over
an inch. Although the spiders vary in appearance,
all orb weavers spin a similar-style web with a spiral
circling outward over lines that look like the spokes of a
bicycle wheel.

A few orb weavers spin their webs parallel to the ground, but the majority construct them vertically. Most are nocturnal and likely to run away if disturbed. At night they come out to survey their web and make any necessary repairs. A few species deconstruct and consume their web every morning, and then spin the entire thing anew at dusk.

TANGLE WEB OR COBWEB (THERIDIIDAE)

Not to be confused with the strands stuck in the corner in your living room, cobweb spiders, from the Theridiidae family, spin messy webs also called tangle webs. They are three-dimensional, unlike the flat webs of orb weavers, but serve the same purpose: to ensnare prey. Although tangle webs don't appear at first to be the work of a meticulous architect, they are complex and constructed in an organized way. In one type, silk threads are rooted to the ground, and when an insect blunders into one, the highly elastic thread snaps and pulls the spider's prey into a dense mass of more silk, thoroughly entangling it and cutting off any hopes of escape.

It takes many species multiple days to build these intricate webs, which are designed with specific prey in mind. Some cobweb spiders lay down threads designed to knock prey out of the air, and others use sticky strands all over the web to ensnare prey. Some species have given up on webs altogether and live a roaming life. A few species are kleptoparasites, living in the webs of other spiders and stealthily stealing their prey.

SHEET WEB (LINYPHIIDAE)

Sheet web spiders belong to the second-largest family in the spider kingdom, Linyphiidae. Despite having a large family, these spiders are mostly tiny and, because of their diminutive size, poorly studied. New species are continually being described.

The sheet web spiders' biggest claim to fame is their skill at ballooning, and because of their often transient lifestyle, they appear in large numbers periodically, only to quickly vanish again. Their second claim to fame is their tendency to create fields covered in gossamer blankets of white. The webs are spun horizontally

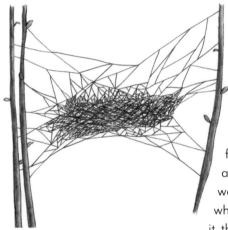

over vegetation, creating what looks like a white sheet. When a large number of spiders spin their individual webs, the effect is spectacular. The webs aren't actually flat but bowl-shaped or hammock-like, with silk threads crisscrossing above, waiting to knock insects out of the air and into the trap below. However, the webs of many in this family aren't sticky like other spiders' webs, and they don't ensnare their prey in the same way. These spiders sit under the web, and when prey falls onto the sheet or walks across it, they bite the invertebrate from below. A few of the tiniest species build their webs in animal footprints. Not all species build webs, however, and, like some cobweb spiders, have taken up the life of a wanderer.

Make a Mushroom Spore Print

It seems unfathomable that the world's largest living organism began life as a microscopic spore invisible to the naked eye. *Armillaria ostoyae*, a honey fungus mushroom, has an expansive network that covers nearly four square miles of forest in Oregon's Blue Mountains.

A single mushroom spore may seem insignificant, but what they lack in size, they more than make up for with quantity. Each day, a single mushroom can release an astonishing one billion spores, as many as thirty thousand every second, which they can keep up for several days. Collectively, mushrooms disperse upward of fifty million tons of spores each year—enough to cover the entire earth to the tune of a thousand spores per square millimeter. Most of those spores are concentrated in forest areas, so

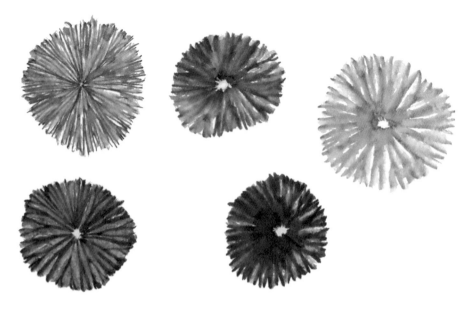

Spore prints

the next time you walk through the woods, think about the fact that you're wading through fungal spores.

With that many spores floating out every day, you might think that mushrooms would not be very particular about how they are released. But many mushrooms have evolved shapes to manipulate the airflow so their spores get caught in the air current and sent to great heights, sometimes even across entire oceans. And they don't even need a breeze to do it.

Mushrooms are classically known to pop out of the ground after a rainstorm. But their relationship with precipitation goes far beyond that. Mushrooms are surprisingly poor at retaining moisture and release it into the air far faster than plants do. But perhaps that's actually a clever adaptation. Scientists found that when mushrooms release moisture, the process of evaporation cools the surrounding air and, in turn, actively moves the air around the mushroom. Mushrooms can create their own breeze. Not only that, but mushrooms create what scientists call a "recirculating eddy," so even barriers growing beneath the mushroom, like plants or other mushrooms, do not pose a problem. The spores simply ride these eddies up and over any barriers. But mushrooms didn't stop evolving there. The shape of a mushroom's cap also influences the airstream around it, similar to how an airplane wing works.

With these adaptations, spores are efficiently sent off into the world, some as high as fifty miles up in the atmosphere. Here, the intimate relationship between mushrooms and rain goes ever deeper. The billions of spores in the air make their presence felt by making it rain and returning the favor to their parent mushroom back on earth. Or so the theory goes.

The intricate relationship of spores and moisture begins with the spore release triggered by what's called the Buller's drop of the mature mushroom. But as the spores move up into the airstream, they dry out. Some spores end up in clouds, along with particles like dust and smoke. We know that those particles form into condensation nuclei, which facilitate the collection of water droplets to themselves. Scientists have shown that in a lab, spores also become condensation nuclei. When the spores are in high-humidity places, like a cloud, they likely reclaim their relationship to moisture by gathering it again. It's entirely possible that mushroom spores actively help make it rain where they are densest, like over forests where other mushrooms are growing. Rain makes mushrooms, and mushrooms—very likely—make rain.

A naturalist can easily study these mighty spores and capture them on paper by creating a spore print. These prints are often natural works of art worthy of display in fine frames, as well as essential for the identification of mushrooms. The caps and gills can vary in color, depending on age and environment, but the spores nearly always retain a consistent color. That's why the first step on a mushroom dichotomous key, the standard guide to species identification, is the spore color.

Differences in spore color are subtle and nuanced, which means that beginners who want to use this characteristic to identify mushrooms face a challenging learning

Warning

Many mushrooms are poisonous, so unless you can confidently identify the species, handle them with care. Always wash your hands thoroughly after touching any mushroom. Additionally, some mushroom spores are known allergens, so if you suffer from seasonal allergies, asthma, or other health problems, you may want to wear a mask when handling them and their spores.

curve. Making spore prints is an excellent first step. And of course, you can learn about spores and mushrooms without diving into the deeper intricacies of identification.

Materials

Knife and Scissors: To collect the mushroom, you'll need a knife. To remove the cap from the stem, you'll want scissors.

Paper: Any type of widely available paper, from copy paper to cardstock, will work. White is standard, but you'll need black paper for mushrooms with white spores. If you want to make artistic prints, you can experiment with other types of decorative paper.

Glass Container: To prevent air disturbance, cover your mushroom with any glass container such as a drinking glass or bowl. You can cover multiple mushrooms with a large container or use smaller jars for individual mushrooms. Make sure the container is small enough to prevent excess airflow and sits flush with the paper when placed upside down.

Artist's Fixative Spray: Spray will fix the spores in place on the paper, preventing them from smudging or otherwise being disturbed. You can find it in art supply stores. Hair spray can work as well.

Microscope, Slides, and Clear Nail Polish (optional): If you want to study the spores under a microscope, you'll need a basic dissecting or compound scope and slides with cover slips. If you want to preserve the spores on a slide permanently, you can use clear nail polish to seal them in.

Steps

Search for Mushrooms: Mushrooms grow in many places, but autumn forests are the most likely place to encounter a wide diversity of species. Search the ground, logs, and stumps. Use a flashlight to spot any that may easily blend in to the substrate. Gilled mushrooms make the most artful prints, but

Tip

You can also buy mushrooms, like portobellos, from the grocery store, if you are uncomfortable collecting wild mushrooms or it is not mushroom season.

nearly all types will release spores on paper, including those with pores. Shelf fungi, the kind that are harder and grow on trees, don't work well for prints, so avoid those.

Determine Spore Color: It's not always possible to see the spore color, but if a mushroom is growing in a cluster, look at the lower mushrooms to see if a powdery substance—the spores from the taller mushrooms—has fallen on their caps. You can also brush your finger gently underneath a cap to see if any spores come off.

Note

Do not collect mushrooms on private land or in parks or other public places where such practices are prohibited. Free or inexpensive mushroom collecting permits are available for many public lands for noncommercial collectors.

Collect Mushrooms: Bring a basket, box, or other hard-sided container to collect mushrooms because they are very fragile and can easily get crushed or destroyed in a pocket or bag. Select the freshest mushrooms, which are mature with open caps but not yet starting to decay. The ideal mushroom is bell shaped with a convex, not flat, or domed top and an even rim. The specimens that make the best prints have an even cap that can sit flush on the paper, but any shape can produce a print.

Use a knife to cut the mushroom anywhere along its stem. Don't pull one from a cluster, especially small ones, or you may pull connected mushrooms out as well. Make note of where and when you found your specimens. Only take a few from each group you find, and if there is only one, leave it and look for mushrooms growing in greater abundance elsewhere.

Prepare Mushrooms: If your collected mushrooms are very wet when you get home, set them out on scrap paper, gill side up, indoors where the air is warm, and allow them to dry out a little. Check them every half hour until they are less soggy, then cut the stems off as close to the cap as possible. Scissors work best because many caps are susceptible to breaking or tearing, and scissors can snip through easily without causing damage. A knife can also work, but it may be tricky to cut through cleanly. Don't try ripping the stem off—doing so can damage the cap.

Prepare Paper: If you have determined the spore color, choose a paper that will allow you to see them easily. For example, if the spores are white, use black paper; if the spores are brown or black, use white. If you don't know the spore color, try half and half. Place half the cap on black paper and half on white, or use an inkpad or black marker to make a section of white paper black. White spores can still be seen on white paper in the right light. Set the paper on a surface where it won't be disturbed.

Place Mushrooms: Place the caps, gill side down, on your paper of choice. You can put as many or as few on a piece of paper as you choose. Space them out enough to be able to cover each with an individual container, or close enough to contain all with one. Place the glass container(s) over the caps and leave them for at least four hours, but preferably closer to twenty-four. Refrain from the temptation to check the spore print. Once you move the cap even slightly, the print will look blurry.

Finish: Remove the glass container, and gently and carefully lift the cap straight up off the paper. The spores will cling to the paper and each other, but it is very easy to smudge them, so don't blow on or touch them yet. To preserve the print, spray it with an artist's fixative outdoors. Be sure to keep the spray at least a foot away from the paper or you may blast away your spores. It's safest to start farther away and move in closer or spray horizontally over the paper.

Let the paper dry outside or in a well-ventilated area and repeat. Once it is dry, dab your finger on a corner of the print, and if any spores come off, spray again. If the spore layer was thick, it may need another layer of fixative spray.

Document: Make a note on the paper of where you found the mushroom, what type it is if you know, the substrate it was growing on, the date collected, when you made the print, and any other relevant information.

> ## Tip
>
> You can make spore prints on a glass pane and even preserve spores indefinitely by taping two pieces of glass together with the spores inside.

> ## Tip
>
> Mushrooms can be hard to remove from the paper without smudging the spores, especially if they're thick. To keep your print pristine, stick a toothpick or paper clip through the tip of the cap to enable it to be easily lifted off. Be careful not to push it too deep as that may disturb the gills.

Study Fungal Invertebrates

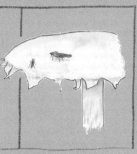

The mushrooms you collected likely came with some invertebrates, which you may discover when making a spore print. When you lift the glass container, you may see tiny specks hopping all over the paper or tiny white grubs crawling around aimlessly. Instead of dumping them outside or squashing them, take the opportunity to study them. A massive diversity of invertebrates live in mushrooms, and you can learn about them as a secondary project. Observe what was living in your mushroom, make notes of what you find, or illustrate them in your nature journal, and then turn them loose outside again, preferably on the mushrooms you found them on.

The tiny, white, wormlike invertebrates with black heads are likely the larvae of fungus gnats and are commonly encountered in mushrooms where the gnats lay their eggs. The hopping invertebrates may be springtails that were hiding in the gills, or they may have hatched as eggs like the gnats. A great many other invertebrates that eat mushrooms may be hiding in the gills like tiny slugs, beetles, mites, and woodlice.

To dive even deeper in your study of fungal invertebrates, you could rear the larvae you find to maturity and learn what they are. Put a layer of sawdust in a glass jar, place the larvae inside, and cover the top with fabric secured with a rubber band. After a few weeks or even months, the larvae may turn into flies in the Phoridae family, known as scuttle flies because they like to run in short bursts. Or perhaps you will see fungus gnats, which include forty-five hundred species of flies in the Mycetophilidae family worldwide.

If you shake or tap the mushroom before making the spore print, you can study the invertebrates first, or you can wait until after you make the spore print to see what falls out. You can also put the cap gill side up under the microscope to see what may still be hiding inside.

Make a Spore Slide (optional): Individual spores are invisible to the naked eye but can be seen under a microscope. Take a sewing needle, razor blade, scalpel, or something else with a fine tip and scrape a few spores off your print.

Tap them onto a microscope slide and place a cover slip over the top. Tape the slip edges to the slide to save the spores, or use clear fingernail polish to seal the cover slip. Be careful not to use too much polish because it could seep under the cover slip. Label the end of the slide with relevant information (type of mushroom, date, location, etc.) about the mushroom.

Common Mushroom Genera

There are dozens of mushroom genera you may encounter. Some are as large as a dinner plate, others as tiny as the head of a straight pin. Here are a couple of common genera you may find.

RUSSULA

A genus that refuses to be overlooked, these mushrooms are often large and bright, with red, purple, white, and other various shades of caps and white stipes, or stems. Although they appear stout, *Russula* mushrooms have a particularly brittle texture, which has led to them being called "brittle gills." Their spores range from white to yellow and ochre. Some are edible, but few of the roughly 750 species around the world are sought by mushroom foragers.

Although it is relatively easy to decide whether a particular mushroom belongs to this genus, identifying them down to the species level is more challenging and frequently requires examination by microscope. *Russula* mushrooms are symbionts with some plants, exchanging resources.

MYCENA

The common name for this genus of decomposers is "fairy bonnets," which gives a good indication of their size. Of the five hundred species that make up *Mycena*, few are more than a couple centimeters tall, but they make up for their diminutive size by being abundant. Although frequently encountered, this genus is largely ignored because mushroom foragers don't eat them, and the tiny fungi are incredibly hard to identify because most require microscopic examination. The classically bell-shaped cap grows atop a daintily slender stem, and most are shades of brown. All species in this genus produce white spore prints, and a few are bioluminescent and glow in the dark.

Grow Slime Molds in a Moist Chamber

SEASON:

Autumn

STUDY TOPICS:

Slime mold sporangia development, diversity

Hidden in the dark depths of the forest grows a minute world of riotous colors and forms. Slime molds, also known as Myxomycetes, may have an unbecoming name, but once a naturalist sets out to know them, they discover a beautiful, tiny world. Gulielma Lister, a British naturalist and slime mold expert, called them *myxies*, a term I have adopted to help bring some respect for these wondrous organisms.

Superficially, slime molds resemble miniature mushrooms. Some are so tiny that they can be seen only under a microscope, though many can be seen without the aid of optics. Myxies form a variety of shapes when they mature, from orbs growing directly on the substrate to tall, cylindrical tubes on slender stalks. They come in a wide range

Metatrichia sp.

of colors, from white to pink and even iridescent, and release spores that aid in identification.

The slime mold's life cycle contains two stages. First comes the plasmodium, where the "slime" in slime mold comes from. Searching tendrils grow under tree bark, over leaves, and even on animal dung, feeding on the bacteria in decomposing organic matter. Unlike tree roots or fungal mycelium, which spread from a central point, slime mold tendrils actively travel in search for food, pulsing and humming as they go. The next, much more glorious stage in the myxie's life cycle is the sporangia, or fruiting body. Some, like *Arcyria* grow tall, oblong tubes that burst open and look like miniature stalks of cotton candy. Others are iridescent orbs, resembling miniature disco balls for the tiny springtails that live among them.

Scientists study slime molds by documenting the fruiting bodies they find in the forest. Because myxies are so small and secretive, even experienced searchers can struggle to find them. Searching for only the fruiting bodies will yield a small fraction of the species in a particular forest. Fortunately, there is another way to explore the diversity and observe the life cycle of slime molds, and it's easy to do at home. A moist chamber is a simple technique that provides insight into Myxomycetes diversity in a particular location, and also offers the opportunity to closely observe the development of sporangia. One of the earliest studies that used a moist chamber happened in 1933, when scientists produced species of slime molds believed at the time to be rare. The cultivation of bark in the chamber revealed that those species were actually quite common and widespread, just rarely encountered because of their diminutive size.

Over time, moist chamber studies have revealed that some Myxomycetes species prefer to grow on the bark of living trees, while others grow on the bark of dead trees, and still others can be found on the bark-free wood of decomposing trees. These studies showed there are differences in types of wood, and they are more complex than collection notes had previously indicated. Bark and wood are not only different physically but also chemically. Bark along with wood even influences the microhabitat of slime molds by preserving moisture and offering protection from potential foragers.

While not a guarantee, a moist chamber is highly likely to produce slime molds. In one study of thirty moist chamber cultures, twenty-nine grew an astonishing thirty-three species of slime molds. Most samples had at least two species, and a few had more than four. This study proved that any Myxomycetes survey should also include a moist chamber technique to get a representative sample of the myxies in any location.

In forests where slime molds can be found, there are dozens of microhabitats, and there is much to discover in sampling wood. The activity that follows affords the naturalist the opportunity to create a collection of Myxomycetes specimens, which are largely lacking in many botanic collections. Few people spend time cultivating slime molds in moist chambers, and there's a wealth of discoveries to be made, so feel free to experiment. You may discover something new!

Materials

Petri Dish: You'll need a large petri dish with a lid. Plastic dishes work best, because if the slime mold grows onto the container, you can cut it apart to preserve the mature specimen. Any clear, shallow container with a lid or cover that can create humid conditions can work, though.

Knife or Screwdriver: You will need a sharp knife to cut a piece of bark or wood from a tree. A flathead screwdriver can also work to remove bark.

Collecting Containers: You'll need something to carry your samples home from the field. Almost anything will work, including bags, envelopes, tins, plastic boxes, and so on.

Paper Towel: To help create humid conditions in your petri dish, you'll need a paper towel to retain water at the bottom.

Distilled Water: Distilled water will keep the chamber humid.

Boxes: If you want to preserve your specimens and keep a slime mold collection, you'll need boxes like craft boxes, matchboxes, or any small cardboard box to store them in.

Steps

Collect Materials: Go into the woods and collect tree bark, leaves you find on the ground or still attached to trees, dung from herbivores (rabbits, deer, etc.), and wood. If you collect a few different things, put each one in a separate container. Take any samples out of plastic as soon as you return home so that they do not get moldy.

Pay attention to the type of bark if you collect it from a living tree. Generally speaking, trees with rough bark tend to produce more slime molds than those with smoother bark. Good examples are maple, oak, ash, and elm. Deciduous trees likewise tend to yield more results than evergreen trees.

Once you choose a tree, remove a piece of bark about the size of a postage stamp from the tree's dead outer layer. You can easily use your fingers to pull away loose bark, but you might need to use a tool like a knife or screwdriver on any that is attached more securely. Try collecting bark and the inner wood from dead trees, including snags and fallen trees.

Make a note of the substrate, type of tree, date, and location you collected your sample from, especially if you're collecting from several different trees.

> ## Warning
>
> Make sure to remove only the dead outer bark. To prevent infections in the tree, do not cut into the living wood tissue.

Prepare the Moist Chamber: Once you have your bark, set up your moist chamber. Cut a paper towel to the size of the petri dish and lay it in the bottom. Next, place your bark or other organic matter on top, spaced so the individual pieces are not touching. Don't combine bark from different trees—put each sample in its own dish—but multiple pieces of bark, wood, or leaves from the same tree can be placed together in one dish.

Cover the bark with distilled water, put the lid on top of the container, and set it aside overnight to allow the bark to absorb the water. Label the container with the type of tree.

The next day, pour off the excess water and set the dish aside where it won't be disturbed. The moist chamber doesn't need any special conditions, just regular room temperature and diffused light.

Monitor: Every day or two for two weeks, check the samples for signs of slime mold growth. Look for the tendrils of the plasmodium or the fruiting bodies. Even if you don't detect anything, keep monitoring the samples every week or so because some slime molds, especially the larger species, can take weeks or even months to develop.

Some of the tiny species visible only with a microscope can develop quickly. If at first glance you don't see anything, try placing the petri dish under a dissecting microscope or use a hand loupe to look closer. If possible, keep the lid or cover on your dish even when viewing it with a microscope—removing the lid may interfere with development.

Add more distilled water when the bark samples begin to dry out; a pipette works well for this task. If you're lucky, you'll end up with multiple species on the same sample.

> ## Note
>
> If after two weeks your chamber is not developing any slime molds, you can soak the samples again in distilled water for twenty-four hours and try again. If after two more weeks there are still no signs of myxies, discard the bark and start fresh or wait longer if you prefer. Remember that some species will take weeks to develop.

Preserve the Slime Molds: If slime molds develop on the bark, you can preserve them. Once they are mature, but before they open to reveal their spores, move the lid slightly to let in a little airflow to prevent mold from growing on the slime mold bodies.

When the myxies begin to open and reveal their spores, remove the bark from the moist chamber. Set it out to air-dry.

After the slime mold and bark are completely dry, place the specimen in a small paper box. Some people like to glue the specimen to the lid, while others glue it to the bottom of the box. If you want to be able to lift the specimen out later to study it or to view it under a microscope, cut a piece of cardstock paper to a size that can be folded into thirds to line the bottom and two sides of the box. The paper should reach the top of the box to keep the slime mold snug and prevent it from moving around and getting damaged. Write your collection notes on the sides, and glue the slime mold in the middle section. You may also want to write the species name, genus, or even just family, if you know it, on the top of the lid so you can organize your collection without having to open boxes.

BONUS ACTIVITY
Keep a Pet Slime Mold

The species you'll observe in this project don't live long in a lab, but there is one species found in labs and artist studios around the world. *Physarum polycephalum* can be purchased from lab supply companies with all the materials you need to keep your pet slime mold going indefinitely as a plasmodium. This species has been used in scientific experiments ranging from navigating mazes to mathematical modeling to creating computers. Artists have also used it to create music.

There is no shortage of activities you can try with your pet slime mold. And all you need to care for it are a few oats.

Your collection notes should include the species, collection location, date, substrate the myxies were on (living or dead bark, wood, leaves, and species of substrate if known), name of collector, and a note that it was cultivated in a moist chamber.

Before adding the box to your collection, freeze it for a couple of days to kill any invertebrates that may be in your sample and could eat and destroy the slime mold. Freezing the specimen also preserves the invertebrates on the sample for future investigation.

Types of Slime Molds

There is no telling what slime mold species you may encounter, but here are a few cosmopolitan species found around the world that may grow in your moist chambers.

ARCYRIA CINEREA

This species is one of the most common in moist chamber studies. The slender sporangia are oblong or cylindrical shapes that grow on top of a tall stalk. Each can sit alone, but they often grow in a cluster with fused stalks, sometimes with upward of twenty individuals. The entire structure maxes out at 4 millimeters tall but can be as short as 0.3 millimeter. It is usually an ashy-gray color but can be brownish yellow as well.

In nature, this species is very common and usually found on dead wood but can also grow on dead leaves and the dung of herbivorous animals.

CRIBRARIA VIOLACEA

This slime mold is an example of a seemingly rare species commonly found in moist chambers that may be much more abundant than field records show. These tiny myxies grow to no more than 2 millimeters high and may be as small as 0.5 millimeter. The sporangia is a deep purple orb with a metallic sheen nodding on top of a hair-thin stalk that is seven times as long as the orb is wide. When mature, the purple orb is covered in a delicate mesh-like structure resembling lace. Similar *Cribraria* species may be seen in a moist chamber but are usually more brownish in color.

In the field, this delicate species has been found growing on the bark of living and dead trees, mosses, and even pig dung.

STEMONITIS NIGRESCENS

This species, along with most *Stemonitis* species, doesn't like growing alone, so look for it in a dense cluster. The individual sporangia consist of a long, tubelike structure attached to a short, very thin stalk and grow between 3 to 5 millimeters tall. As a cluster, it looks like a little slime mold bouquet. This species is black, a key trait that distinguishes it from the similar *S. fusca*. Until recently, these two species were considered the same species and *S. nigrescens* was believed to be a shorter, darker variation of *S. fusca*. It is, however, not a variation but an entirely different species.

Out in the field, they can grow in large colonies on dead wood and bark but can also be found on decaying plant matter like leaves.

COLLARIA ARCYRIONEMA

This species glitters and glistens like minute jewelry. Its diminutive size of only 1 to 2 millimeters high does nothing to negate the glamour of these myxies. These glittering, iridescent metallic orbs range from gold and silver to bronze and blue. The peridium, or outer skin, flakes off to reveal a brown interior. The slender stalk accounts for about two-thirds of its height, and they grow singly but scattered in

large aggregations. Until very recently, this species was known as *Lampro-derma arcyrionema*.

In the wild, these myxies grow most commonly on dead plant matter, particularly leaves, but can also be found on wood and animal dung.

Make Tree Bark Rubbings and Casts

SEASONS:
All

STUDY TOPICS:
Tree bark anatomy, microhabitats

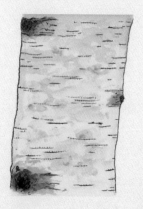

While the rings inside a tree may tell us how old it is, studying its bark can reveal a tree's life story. A naturalist could spend a lifetime learning one tree's secrets.

Bark is never static. It changes slowly as it ages and is also altered by the environment it is rooted in. Fire, lightning, fungal infection, bacteria, frost, ultraviolet rays, parasites, insects, and other animals—they all leave their mark, marks that a keen-eyed naturalist can deduce with practice. Bark can also tell the story of the tree's surrounding environment. The presence or absence of lichens, mosses, and other epiphytes (organisms that grow on plants) can tell a naturalist if pollution or other anthropomorphic influences are at work.

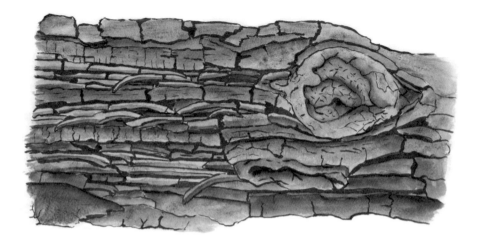

Bark of Pacific madrone

Bark is a tree's suit of armor. And trees have developed a staggering variety to protect their most vital living tissue. Like armor, bark protects against a range of threats, from climate and weather to natural disasters, fungi and bacteria, and the many animals that want to consume the wood inside.

Trees evolve their particular suits of armor in response to their environment. Some bark is smooth and thin, some is thick and deeply fissured, while yet other armor is papery or fibrous. Where fire is common, bark may be fire resistant or may flake off and pile up on the ground so wildfire burns hotly through, bypassing the more vulnerable trunk. White or silver bark can help reflect the sun's ultraviolet rays in cooler climates.

Bark consists of several layers, called periderm. The cork, or the outermost surface that most people think of as bark, is really just the skin. Underneath are the layers vital to transporting nutrients and water. Just inside the cork layer is the secondary phloem, or inner bark, which continues to grow to replace any bark that falls off and retain a uniform thickness. Dividing the bark from the actual wood, called the secondary xylem, is the hardworking vascular cambium layer, which produces cells for both sides. New layers are continually being added inside, and the outer dead cork layer that we see serves as a place to shunt waste materials in the form of tannins, gums, resins, or crystals.

Time passes quite differently for trees compared to humans. You may feel like you're well acquainted with particular trees, but even if you visit them regularly over years, even decades, you will never know a tree on its own time. Some of the oldest trees in the world can live thousands of years, and their growth is often so gradual we don't even register it.

As a tree grows, its bark slowly changes, as do the communities of organisms the tree hosts. Moss and lichens come and go with the changing of the bark and the invertebrates that live in those cryptogams with them. The bark of mature trees can be covered by a complex topography that hosts miniature worlds. Spiders make webs in the crevasses, while other invertebrates run around in vertical chasms. Hidden in moss and lichens, tardigrades, springtails, and rotifers live in microscopic drama. Birds, in turn, hunt along the trunk, hoping to feast on those invertebrates. Bats find cozy roosts under loose chunks of bark, and mammals like beavers and porcupines feed on the bark. Other mammals like deer use the rough bark to rub the velvet off their antlers, and bears use it as a back scratching post.

Humans have also made use of different types of tree bark. Many Indigenous tribes of the Pacific Northwest use the bark of the western red cedar (*Thuja plicata*) for items including ropes and mats. Cultures around the world have written on birch bark, with samples dating as far back as the first century CE. In Asia, mulberry bark was harvested to create paper, a technique first developed in China as early perhaps as the second century CE. Some medicines have even been developed from bark. The first aspirin, for example, came from willow and poplar bark before manufacturing shifted to synthetic materials, and bark from the quinine tree was an early treatment for malaria. We still use products made from tree bark like cork, rubber, incense, and cinnamon in the twenty-first century.

Exploring tree bark is a sensory experience. The gums, resins, tannins, and other substances give bark from different families distinct scents. The texture varies widely, and touching tree bark is a journey that ranges from silky smooth birch to fibrous western red cedar to the painful spines of hawthorns. Bark also grows in many patterns and colors—what a visual treat for the naturalist.

Tree bark provides wonderful opportunities for a naturalist to become an observer, seeing things most everyone else's eyes simply pass over. We can truly appreciate this living art gallery that is waiting for us, right outside our doors.

Leaves, seeds, and even small stems are easy to glue or tape to paper and preserve in an herbarium, but the thickness of most types of bark presents a logistical challenge. Additionally, collecting leaves, twigs, or cones does not harm a tree, but cutting bark off a living tree can create a vulnerable wound.

However, there are a few ways to document and record tree bark. The simplest is, of course, picking up pieces of bark from the ground and storing them in boxes. But because the fissures and other texture of some mature bark can be so wide and deep, it needs to be represented in larger pieces than most people can reasonably store.

Some naturalists take photos or draw illustrations of tree bark, but two other methods are reliable options for documenting bark. First is the very simple but tried-and-true method of bark rubbings. The second is a little more involved, but with the most accurate representation: tree bark casts.

Bark Rubbing Materials

Paper: Depending on your goals, you have many options. Simple copy paper works well and is a uniform size, but some mature trees may be too large to capture a good representation of bark on letter-size paper. You can also cut kraft paper to the desired size to capture a particular tree's bark. Its brown color also suits tree bark rubbings well.

Pastels, Crayons, or Charcoal: Wax crayons are the most common tool to make bark rubbings, but other art supplies can work too. Pastels come in a variety of colors and have a more delicate touch than crayons. Charcoal is soft and works on different types of bark. Whatever you choose, select a tool that is thick. Thin pastel or charcoal sticks will break easily. Longer sticks work better than short ones, because the shorter your tool is, the less area it will cover.

Tape: You can use masking, packing, or other heavy-duty tape, but be prepared to cut the edges off because trying to remove it would rip the paper. A second option is to find specialty artist or drafting tape, which is designed to be removed from paper without ripping or tearing but will still stick to the bark.

Artist's Fixative Spray: If you use pastels or charcoal, you'll need an artist's fixative spray to prevent the rubbings from coming off, smearing, or creating a mess and destroying the rubbing.

Bark Casting Materials

Clay: To make a mold, you'll need artist's modeling clay or plasticine. Look for nonhardening clay, which doesn't harden in the air. It stays soft and can be reused for more casts if it's kept clean enough.

Plaster of Paris: Easy to find at any hardware store, plaster of Paris consists of a box of very fine powder that you mix with water and pour into a mold.

Board and Hammer: You can press the clay into the tree by hand, but putting a board over the bark and hammering it will push the clay into the cracks and crevices. Any type of wooden board and hammer will work.

Sand: You'll need a bag of fine sand to nestle the curve of the mold.

Box: Your box should be made of cardboard, like a repurposed shoebox, and slightly larger than the mold you create. It will confine the sand, mold, and any plaster that may spill. You'll need to pour sand along the sides once the mold is inside, so make sure there's room around the edges.

Paint and Glue (optional): Once you have your plaster cast, you can paint or glaze the white material to replicate the tree bark. Applying a layer of diluted white liquid glue will help seal the plaster and make it less likely to absorb the paint.

Steps: Bark Rubbing Method

Find a Tree: Pick a tree with interesting bark that is safely accessible. Try many different types, and look for trees with different textures of bark on the trunk and branches. Survey the tree for any danger, like wasp nests, before proceeding.

Prepare the Paper: Find an area on the tree with representative bark. Look for a location without rough knots or other

Note

Observe sound ethical practices, and don't trespass on private property or stomp on sensitive plants to get to a tree. Don't disturb any wildlife that may be in the tree.

protrusions that will prevent the paper from lying relatively flat. Avoid mosses, spiderwebs, or anything else living on the bark that could get damaged.

Tape the paper to your chosen place. For a small sheet of paper, taping the corners may be sufficient, but to keep a larger sheet of paper in place, you may need to tape all four sides. The slightest shift can ruin the entire rubbing.

Rub: Using the long side of your crayon, pastel, or charcoal, slowly rub along the bark. Start gently, and depending on the bark, rub more firmly. Watch out for the ends of your pastel or charcoal and crevices in the bark, which could poke a hole in the paper. Also be mindful of the ends of your tool, which may create unwanted streaks. Making a good rubbing can take practice and patience, so be sure you have enough supplies to try again if you need to.

> ## Tip
>
> Carve or sand any angular edges off the ends of your pastel or charcoal stick to keep from making lines on your paper or creating holes.

Preserve: If you are using pastels or charcoal, once you are finished with a rubbing, spray a layer of artist's fixative over the paper. Once it's dry, rub a corner with your finger, and if a lot still comes off, spray another layer. You may need to apply several layers. The fumes from the spray can be unpleasant, so leave the papers outside or in a well-ventilated area until they're dry.

Steps: Bark Casting Method

Find a Tree: Follow the guidelines in the bark rubbing steps to find a tree for your casting. Take a photo of it to reference later and replicate the colors if you decide to paint the finished cast.

Prepare the Clay: You can prepare your clay at home by warming it slightly in the microwave and rolling it out into a square or rectangle so that it's ready to press into the bark. It's possible to roll the clay in your hands in the field and simply press it into the bark, but some clay can be really difficult to flatten by hand, especially when cold.

The thickness of the clay will depend on the bark. Thinner clay will work for trees with less texture, like birch or alder, but oaks, pines, and anything with deep texture need somewhat thicker clay. The clay needs to be thick enough to reach into all of the bark's grooves while retaining a flat and unbroken outer surface that prevents the bark from poking through.

Press: Follow the guidelines in the previous method's steps to find the right place on the tree you've selected. Firmly press the clay into the bark. Take your board and hold it over one area of the clay and hammer on it. Work around the clay until you've covered the entire area with the board and hammer and the clay has reached all the depths of the bark. If you are not thorough, the cast may end up with flat areas, so go over it a couple of times.

Slowly peel the clay from the bark, being careful not to press your fingers into the textured side. The clay will lose the curved shape of the trunk but should still reveal all the indentations and ridges of the bark.

Clean the Clay: Some bits of bark and other debris from the tree may stick to the clay. Pick them out with your fingers or tweezers. The cleaner you can get it, the better, but little bits of bark and debris are fine.

Prepare the Mold: Back at home, pour a thick layer of sand into the box. Gently curve the clay mold so the bark texture curves inward to once again represent the natural curve of the tree trunk. Then, build a wall of clay on each end of the mold, leaving the back open. These will act as dams to keep the liquid plaster mix in the shape of the tree, instead of running out into the box. Keep one end as flat as possible so the mold can eventually stand upright on a flat surface.

Nestle the clay mold into the sand and then pour more sand on the sides so that it supports the curve of the mold and won't pull away from the walls on the ends or be flattened by the weight of the plaster.

Prepare the Plaster: Follow the instructions on your package of plaster of Paris to get the proper ratios of powder to water. Slowly pour it into your mold until it's full. Let it set. Consult the plaster of Paris instructions for time frames, usually about twenty minutes to a half hour.

Finish: Once the plaster is hard and dry, remove the whole thing from the box and gently peel the clay off. You can remove any

> ## Tip
>
> Sand will fall off when you remove the cast, so do this part somewhere you don't mind getting messy, like a basement, a garage, or outside. You can also set the box on a tray to contain the mess.

stray bits of clay with a soft brush, like an old toothbrush. If the end isn't flat enough to stand upright, you can sand it down. Once the cast is all cleaned up, you're ready to paint it.

Types of Bark

There is no shortage of tree bark nearly anywhere a naturalist lives. Not only do trees have a variety of textures to document, but each tree has different textures from the trunk to branches. And each species will vary slightly depending on age and location. Here are two common types you may encounter.

OAK (QUERCUS SPP.)

With about five hundred species of oak in the Northern Hemisphere, there's a good chance you'll find one nearby. Oaks have tremendous cultural, economic, and ecological importance. Their wood is very strong and has been used in everything from Viking longships to the House of Commons in London. An oak tree appears on coats-of-arms from Oakland, California, to Norway and is a tree of massive importance in folklore around the world. It is even more important ecologically because it's a keystone species, shaping the entire habitat where they grow. Hundreds of species of insects depend on oak trees, and many birds and mammals rely on the acorns produced by the trees.

Most oak bark comes in shades of brown or gray, and mature trees are usually deeply fissured or furrowed. There are always exceptions, and a few species of oak have smoother bark. Young bark is often smoother, as well, compared to older trees. Red oak bark has more vertical fissures, while white oaks form bark that is generally more blocky.

BIRCH (BETULA SPP.)

As smooth as the oak is rough, birch trees are famed for their crisp white bark. However, there are multiple types of birch trees, and some species are not smooth and white at all, but brown and gray and roughly textured. Birch trees often vary dramatically between young and old trees. Young white birch species are smooth but often dark red or reddish brown in color. Mature white birch species have hor-

izontally lined, white bark, often furrowed to reveal a dark wood underneath. Some mature trees shift the orientation of their bark pattern, with fissured plates growing vertically. The darker bark birch species also change over time. Young trees have brown, reddish, or gray bark with horizontal lines. While the bark of some white birch species peel, others do not. As tempting as it is, never peel the bark of a birch tree because that can kill it.

Birch are important pioneer species, meaning they are among the first trees to sprout and grow when there's a disturbance, like a wildfire. Like many pioneer species, they tend to be fairly short-lived. Birch also hosts hundreds of insects that feed on the tree, and many other wildlife species, from birds to mammals, benefit from birch trees. The birch also has cultural and economic significance. The Celts used birch in the Ogham, their tree alphabet, and it's the national tree of Finland. Some Indigenous tribes use birch bark to make canoes.

Press Plants

SEASONS:

Spring, summer

STUDY TOPICS:

Plant anatomy, plant and insect interactions

Perhaps no other practice has introduced more potential naturalists to natural history than the time-honored tradition of pressing plants. Early European explorers traveled the world over in search of new plants, sending the majority back as pressed specimens, but it was the Victorians who really made botany a common household activity. Plant identification and pressing became acceptable activities for amateur botanists and especially women, because it was seen as genteel science.

Amateur botanists bought plant identification books and mastered the scientific names, then compared notes and specimens at newly formed clubs. Some became experts in certain plant families or started creating art and became botanic artists.

Plant pressing, a valuable naturalist's tool

Many women who started with botany soon discovered the much larger world of natural history and became fully fledged naturalists. Although they enjoyed significantly less recognition for their work (and still do today), they were suddenly free to go outdoors to tromp around in the mud, climb cliffs, and botanize. It offered a freedom largely denied to women at the time.

Pressed plants are still an important tool for naturalists today. In addition to helping with identification and the study of plants, they retain a great deal of sensory information lacking from photos and drawings. With a pressed plant, a naturalist can feel the texture and know whether a leaf is fuzzy, smooth, or prickly. They can catch the scent of the dried plant. And although the colors will fade some, naturalists can see what a specific plant actually looked like in person.

Because the field of botany is immense, naturalists can investigate a variety of topics with their pressings. They may choose to collect and press plants to learn more about a certain family or to identify tricky groups. Perhaps they want to identify and learn about an unfamiliar plant they encountered during a walk or create a collection of plants that represent a particular location.

One of the things that makes being a naturalist unique is that we're not limited to finding a perfect specimen, and we can focus on plants that tell a complete story. We may preserve and document the interactions of plants and other organisms, like the fungi or cryptogams (moss, lichen, and liverwort, for instance) that grow on plants. Or leaves nibbled on by a caterpillar, wasp galls on leaves, stem scars from dragonflies laying eggs, cones chewed up by squirrels—you may find any or all of these of special interest. A naturalist has the freedom to pay attention to whatever catches their interest and decide how to collect, preserve, document, and display specimens.

In addition to pressing plants, you can dry them or preserve them in a frame, vase, or box. For example, you can make a display that illustrates the life cycle of a plant by including everything from seed to flower in a jar, box, or frame. Naturalists can make prints directly from plants or press them into sculpting clay, which they can then bake and paint. Or naturalists can make plaster casts of plant material. (See Project 5 for instructions on making a leaf cast from plaster.)

There is an infinite number of ways to display the story of a plant. For a personal collection, the only limit is your imagination.

Materials

Collecting Tools: It's good to have a few tools in your field bag at all times for collecting plants, whether you're searching for plants or just happen upon something unexpectedly. These are a few handy tools for collecting plants:

- A metal, fold-up shovel to help dig up plants and roots you wish to preserve
- Clippers or a knife for collecting twigs, branches, and other woody material
- A hand lens to help you note fine anatomy details for identification or observation purposes
- Plastic bags to carry specimens that you plan to press after you return home; a bag will prevent them from drying out and wilting
- A pocket press if you wish to press small plants in the field immediately (see the bonus activity for instructions to make one)

Make a Pocket Press

Lightweight, compact, and made largely from repurposed materials, a pocket press is the perfect tool for a naturalist's bag. You'll need some corrugated cardboard, duct tape at least two inches wide, kraft paper or newsprint, and a rubber band.

Start by deciding on the size of your press. A four-by-three-inch press is generally small enough to stash away in a bag but large enough to press flowers, leaves, and other small plants. Cut between ten and twelve pieces of cardboard to the size of your press. The more pieces of cardboard, the more room you'll have for plants, but your press will be bulkier.

Cut sheets of newspaper or kraft paper to place between the pieces of cardboard to help secure the plant in the press and remove moisture. The pieces of paper should be the same height as the cardboard but twice as wide. For example, if the cardboard is four by three inches, cut the paper to four by six inches, then fold it in half like a book.

Assemble the press like an accordion. Begin by cutting a piece of duct tape twice as long as the longer edge of the cardboard, plus another inch. If your press is four by three inches, cut the tape nine inches long. The tape will act like a flexible book binding. Lay the tape down on a flat surface and visually divide the width into thirds—it doesn't have to be exact. Place a piece of cardboard on the tape, covering the right third. Place another piece of cardboard in line with the first piece on the left third of the tape, leaving a gap in the middle. Finally, fold the duct tape over, sealing the long edges of cardboard together and covering the exposed sticky gap.

Repeat this process for the rest of the cardboard pieces with more pieces of tape, connecting one piece to the next until the press is one long line of cardboard and tape. Then fold the whole thing accordion-style. Place the folded kraft paper between each piece of cardboard, open side facing the spine. Keep the whole thing together with a rubber band.

As you place plants in the press, make notes on the kraft paper about the location, date, plant name, and other relevant information so you don't forget later.

Plant Press: You can buy a plant press, but it's also possible to construct your own with easy-to-find materials and wood, following instructions that are easy to find online. Standard-size plant presses are 12 by 18 inches, but many smaller sizes work just as well. The main components of a plant press are a wood frame with some type of binding, like bolts or a strap. Inside the frame are pieces of corrugated cardboard, which help press while allowing air to circulate.

Glue: While you can buy specialty herbarium glue, standard white liquid glue found in any art or office supply store will also work.

Paper: You'll need several types of paper to press plants. The first is a blotting paper to absorb moisture from the plant and help it dry. Newsprint is most commonly used because it's the perfect size, but kraft paper works well if you cut it. Choose or cut your paper so that when folded in half like a book it is the same size as the press. Botanists often use special blotting paper, which is available online.

> ## Tip
>
> It is also possible to collect fungi, mosses, lichens, liverworts, and other epiphytes that grow plants, but there are special methods for collecting, preserving, and storing them, so consult how-to books for instructions or search online.

Second, you'll need paper to mount your specimens after they're pressed. Specialty herbarium paper is available online, but you can choose from a wide range of other papers for your own collection. The standard size is 29 by 42 centimeters, but small samples can be mounted on smaller paper and even colored paper. Some naturalists prefer to make a book of specimens or mount them directly into their nature journal. The only criteria is to choose paper that is stiff, not thin like copy paper.

You can use separate, nonacidic, basic white paper to make labels for your specimens, or you can simply write on the paper the plant is mounted on. The standard label size is 11 by 7 centimeters.

Paintbrushes: These brushes are for applying glue, so use inexpensive ones. It helps to have two sizes: a wider, thicker brush to distribute the glue and a fine brush to touch up parts of the plant you may have missed or that are loose after the plant was mounted. White glue can be washed out of the brushes so that you can reuse them.

Wax Paper: Simple wax paper from a grocery store works fine for spreading glue to lay the plant on.

Forceps or Tweezers: Many plants are too delicate to manage with your fingers, so you'll need a pair of forceps or tweezers.

Weights (optional): To hold some plants down while gluing, you may need weights like metal washers, fishing weights, or other small but heavy things.

Paint and Brayer: To make prints, you'll need basic paint and a small hand roller, called a brayer. Both are easy to find in art supply stores. For a simpler method for small specimens, you could also use a basic stamp pad.

Steps: Pressing Plants

Collect Plants: You may be out specifically looking for plants to press or just happen upon something interesting while you're in the field. You can document leaf variations or entire plants by collecting leaves, flowers, fruit, seeds, bark, twigs, and any other accessible parts. You could also visit the same plant throughout the year to collect the entire life cycle.

If the plants or parts are small, you can put them directly into your pocket press, using the directions below. Otherwise, put them in your bag or other container to keep them safe and prevent them from drying out too quickly. If you're collecting plants from different locations, put a tag on each one with relevant information. If you want to press the roots or bulb along with the plant, use a knife or plant clippers to cut the stems. Don't rip pieces off, which can damage the plant.

> **Warning**
>
> Become familiar with poisonous plants where you are. Avoid any that may cause skin reactions, or handle them with great care.

While most botany sources recommend making pressings of perfect specimens, you can get a more complete story if you also collect some with damage. If you see an insect changing a leaf in some way, document that with a photo or a sketch, and make notes in your nature journal.

Try to avoid collecting when it's raining, because wet plants may become moldy before they dry. Be sure to collect ethically. That means knowing the rules where you are and honoring them. Rules vary widely from country to country and within different countries and even cities and parks. In addition to following local rules, also follow sound naturalist ethics. Don't collect rare or endangered plants, don't take more than you need, and don't destroy habitat to get to a plant.

Document: Write down relevant information while you're still in the field, especially if you want to create a herbarium or plan to donate your collection one day. The common data on standard herbarium labels includes: species name (scientific and common), location (as specific as possible), date collected, habitat type, collector's name, name of person identifying the plant if different from the collector, and any other pertinent information.

Pressed lingonberry

Press Plants: At home, if your specimens are not already in a pocket press, prepare them by cleaning off any dirt and other debris. Transfer your observation notes to the blotting paper for each individual plant. Unfold the paper, and place the plant on one side.

Carefully arrange the plant, unfolding leaves and spreading it out so the relevant parts are visible once pressed. When pressing flowers, arrange the petals so the inside parts or other important characteristics are visible. Consider turning one leaf over to show the underside. If the plant won't lay flat as you work, use weights to hold parts of it down. If the plant is longer than the paper, a standard method is to fold it in a V shape, and if that's not enough, an N shape. This step may seem tedious, but it's important, and there is no undoing it, so take care with this process.

Once you're satisfied, fold the other half of the blotting paper over the plant, place it carefully between two cardboard pieces, and put it in the press. If you

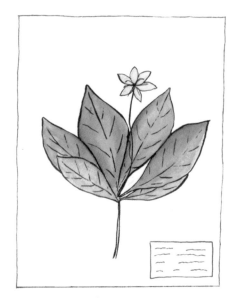

Pressed Arctic starflower

are pressing multiple plants at once, repeat the process. Tighten the press once all the plants are in place. It needs to be quite snug to prevent leaves and petals from getting wrinkly.

Leave the press for twenty-four hours in a warm space with good air circulation. If you're using a lighter-weight press or pocket press, you may be able to open it after a few hours and adjust the plants a little, unfolding leaves or straightening out parts. Just be careful—if the plant is very dry, it can become brittle and break.

After twenty-four hours, open the press and replace the blotting paper for each plant. Plants preserve better when dried faster, so changing the paper will help retain more color, especially in flowers. Put all the plants back in the press, tighten once more, and leave it alone again. It depends greatly on the plant, but most will dry within a few days. Others may take up to a week or more. Once dry, remove them from the press, but keep them in the blotting paper until you are ready to tackle the next step.

Mount: Once the plants are dry, they're ready to be mounted. Spread out a piece of wax paper that's larger than your plants. Find a container, like a glass jar, or repurpose a plastic food container, to mix some white glue with a few drops of water. Mix it together with your brush. There's no specific ratio of glue to water, so working with your particular glue will require trial and error. Practice first on specimens that aren't your best. The glue should be thinner than when it comes out of the bottle but not overly watery.

Spread the glue mix on the wax paper and carefully lay the back of the plant on it. Take extra care because small, delicate petals can become stuck. Use a fine brush if necessary to add small amounts to the most delicate parts. Gently tap parts down so they get an even layer of glue. Carefully pick up the plant with forceps or tweezers and position it above the mounting paper. Then gently place it down on the paper.

If the plant is large, weight it down by placing a piece of wax paper on it and then laying something over that. You don't need much weight. Alternatively, you can use weights, like wide metal washers or other heavy items, to press down on strategic places on the plant.

To document your specimen, add a label with all the relevant information on the mounting paper. If you plan to donate your specimens to a museum or herbarium someday, put the label on the bottom right corner, the standard location in herbariums. If it's for your own personal collection, you can place it wherever you like or write directly on the paper. Let it dry for a day, and then admire your finished specimen.

Steps: Plant Printing

Collect Plants: First follow the previous steps for collecting and documenting plants. You can use fresh or dry leaves and plant material, but dried plants generally work better for printing.

Ink Plants: Place paint onto a piece of wax paper and roll the brayer through the paint to coat it thoroughly and smooth out the paint. To ink both sides of the leaf or plant, place it on the paint and then roll the brayer over it. Then place the leaf on a piece of paper and fold the paper over it. Roll a clean brayer over it to press the ink onto the paper and get a print of both sides on one piece of paper.

To ink just one side, place the leaf on clean wax paper and roll the brayer with paint over it. Place the leaf paint side down on paper, cover it with a piece of wax paper, and roll a clean brayer over it to make a single print.

Grow a Microbial Garden

SEASONS:

All

STUDY TOPICS:

Bacterial interactions, microbial habitats

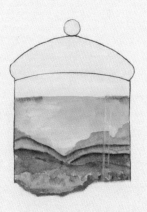

onsider the legends you've heard about giants. Stories from around the world tell of humanlike beings thirty feet tall, lumbering around, unknowingly flattening houses while eating cattle whole. These tales reflect the human perspective that we are "normal" sized, so giants are large and insects are small. But what if that perspective is all wrong? What if we humans are actually the giants and "small" is normal?

The vast majority of life on earth is what we'd consider small. There are only about 6,400 species of mammals in the world, but more than 160,000 known species of flies alone (and likely several times that number still undescribed), not to mention the rest of the invertebrate kingdom. By the numbers, we are the giants flattening the houses of other beings.

When it comes to bacteria, we are truly behemoths. The largest bacteria species, at a whopping three-quarters of a millimeter, can barely be seen with the unaided human eye. Most are microscopic in size, and it's only because many are colorful—when they grow in large swaths of purple, green, orange, red, or black—that naturalist "giants" can still perceive them without any visual aids and even learn to identify them.

Bacteria doesn't live in a vacuum and is a complex part of our ecosystem. Wherever there is moisture, bacteria can be found, including inside our bodies.

And that moisture doesn't need to be large bodies of water like ponds or creeks—it could be the dew on a leaf or the water in animal dung. Bacteria can be found in a variety of habitats, from ponds and streams to cliffs, forests, and even urban environments. Many species are associated with specific habitats just like any other group of organisms.

There are about two thousand described species of bacteria, but as with most "small" things in this world, the actual number of species is likely several times higher. As we've seen, the few species that have a negative effect on humanity get most of the attention, but the pathogens among the bacteria account for a tiny percentage of all known species.

Back in the 1880s, bacteria was the catalyst for scientific debate and innovation. German microbiologist Julius Richard Petri invented, as you could probably guess, the petri dish, a tool designed to grow individual species of bacteria in what is called a "pure culture," untainted by any other organisms. It became a tool used in science labs around the world and is still the primary method for studying bacteria. Russian microbiologist Sergei N. Wino-

A Winogradsky column

gradsky was, however, critical of this technique.

Winogradsky argued against the pure culture petri dish because, he said, bacteria never live alone. They are part of not only the larger ecosystem where they live but also a complex microbial ecosystem. He wanted to understand their ecosystem and how they interact with one another as well as the larger environment. To that end, Winogradsky created his own invention, a bacterial garden, a mixed culture opposed to a pure culture. Despite its many benefits, the Winogradsky column, as it is known today, has been forever overshadowed by Petri's pure culture dish.

A Winogradsky column offers naturalists and scientists alike the opportunity to study microbes, specifically bacteria, and their abundance, variance, interdependency, and general diversity. It's also a chance to observe the interaction of a microbial ecosystem in a microcosm.

All life on Earth falls into four categories when it comes to receiving carbon and energy: Heterotrophs, like humans, gather energy from eating plants and animals. Phototrophs, such as plants and algae, gather energy from light. Chemotrophs, like microbes living on deep-sea thermal vents, get their energy from chemical reactions. And finally, autotrophs, also known as primary producers, such as lichens and coral, produce their own food from carbon sources and energy from light. Incredibly, bacteria are the only group that practice all four of these life strategies, all of which can take place in a single Winogradsky column.

As a bacterial garden grows over weeks, months, and even years, it will develop environmental gradients, whose different colors are easy to see, and a naturalist can observe the boom-and-bust cycles. Although the system is closed, it doesn't die, because the nutrients are recycled and reused as long as the garden exists. It's a self-enclosed, living microbial garden, just like the wetland in a jar (Project 3).

If you've brewed beer, made sourdough bread, or pickled anything, you've already cultured bacteria. The naturalist can easily learn more about microbes and keep a surprisingly beautiful garden on display in their house—talk about a conversation starter.

Materials

Clear Cylinder: When choosing a container, bear in mind that a microbial garden will live for months or even years. There is no standard sized container. Some are as small as a mason jar, while one in a particular museum is the height of a room.

The container should be clear, taller than it is wide, and have smooth sides. You will need to find a cover, either a loose lid or simply plastic wrap secured with a rubber band. Repurposed plastic soda jugs, tennis ball containers, pasta sauce jars, and large mason jars can all work. Or you can look for something more decorative. Thrift stores are great places to find tall glass containers.

Mud and Water: The base of your garden will be mud and water from a freshwater source, and the bacteria contained in it. Look for a water body with muddy sediment like a lake, pond, wetland, marsh, stream, or estuary.

Shovel, Jar or Bucket, and a Bottle: You'll need a shovel to scoop up the mud, and a jar or bucket to carry it home, depending on the size of the container you select. An empty water bottle works well to collect water.

Bowl: You'll need a bowl large enough to accommodate all of the materials to mix together.

Source of Carbon: There are many easy options for adding carbon to the microbial garden. Some will release it quickly, while others are slower. It doesn't matter which you choose, and if you want to, you can experiment. You can choose a single source, a selection, or experiment with different sources in different microbial gardens. Among the fast-release options are baking soda (sodium bicarbonate), chalk, garden lime or crushed oyster shells (calcium carbonate), and antacids. Slow-release options include shredded newspaper, egg shells, and dry organic materials like leaves, sawdust, grass, oatmeal, or cornstarch.

Source of Sulfur: Options for sulfur include Epsom salts, plaster of Paris, gypsum (calcium sulfate), or egg yolk. The amounts will vary depending on the size of your container, but you only need a small amount.

Rubber Gloves: It's a good idea to wear rubber gloves, like dishwashing or medical gloves, to protect not only from the mud but also from any existing bacteria.

Wooden Spoon: To mix the microbial garden, you'll need a wooden spoon, paint stirrer, or some other tool that you don't cook with because it will get muddy and exposed to bacteria.

Diatomaceous Earth (optional): You can find this silica-rich compound in hardware or gardening stores. Add this to your mixture to make the soil lighter in color, which can make it easier to see the bacteria. It can also help fill up space if you don't have enough mud for your container. An additional bonus is that it provides a new, untouched surface for bacteria to colonize.

Steps

Collect Mud and Water: Use a shovel to scoop out mud from your chosen source to fill most of your container (about 85 percent). Try to avoid collecting debris like sticks and rocks. Once you have enough mud, fill the water bottle with water from the same source, again avoiding debris.

If you're not going to prepare your garden right away, store your materials in a shady place until you're ready for them.

Mix: Dump your mud into the mixing bowl and begin to stir it with either a large wooden spoon, a paint stirrer, or your gloved hands. Pick out any large debris like stones, branches, leaves, or garbage. Pour in some of the source water and continue stirring, adding more water until the mud becomes the consistency of a milkshake. If using diatomaceous earth, mix it in next until the soil is the color you want, adding water to retain the milkshake consistency.

Add in a handful of the carbon and sulfur sources (you don't need much), shredding large things like leaves or newspaper first. Mix until everything is thoroughly blended together.

Pour: Carefully scoop or funnel the mud mixture into your container until it's about 85 percent full, going slowly and making sure to pack it down as you go to avoid air gaps. Slowly pour in the source water, leaving a quarter inch of air at the top. Put on a loose-fitting lid, or cover with plastic wrap and a rubber band to keep in any odors.

Place: The microbial garden needs light, but how much will require some experimentation. You can put it directly in a window or farther away but still relying on indirect natural light, or you can use artificial light sources. The bacteria's growth will vary on different sides, depending on the light exposure.

Wait and Observe: This project requires patience. For the first few weeks, the Winogradsky column will look like nothing more than mud in a jar. Microbial gardens grow slowly, and changes can be gradual. Around the two-week mark, you

Warning

Do not seal the container tightly, because the gases the garden releases may explode under pressure. Also don't breathe directly over the column to avoid inhaling any noxious gases.

should begin to see some hints of green in the mud. If you don't, the column may not be getting enough light. Somewhere around four to eight weeks, bacteria may begin showing up. Patches of the same color are likely made up of the same type of bacteria. Eventually the gradients that show up will begin to tell a story. What is growing closer to the surface and the available oxygen? Which bacteria are growing on the dark side, and which are growing only in the light? Which bacteria have other types growing alongside them?

Keep an eye out for any invertebrates you inadvertently collected with the mud. You may see worms, snails, fly larvae, or other aquatic invertebrates in the mud or water on the top.

Make Notes: In your nature journal or another place you like to take notes, record what you added to the Winogradsky column. Continue monitoring it, and consider taking photos or making notes every week or month once you notice bacteria beginning to show up to document the ongoing changes.

Repeat: Consider making several microbial gardens to experiment with. If you collect enough mud and water, you can make more than one from the same source, and then vary the carbon and sulfur materials or vary the amounts. You can also try putting them near different light sources. Label the individual gardens carefully, and track differences in bacterial growth.

Another option is to make different gardens out of different mud and water sources. Look for sources from extreme conditions, like hot springs, as well as the usual fresh and saltwater sources. If you have access to a marine habitat, you can culture different bacteria with sand, seaweed, and other tidal materials.

Some bacteria like metal, so consider adding nails or other pieces of metal to your garden to see if different species grow, and gauge how the garden reacts.

Types of Bacteria

Your mud sample will likely carry many types of bacteria, some of which will become apparent over the weeks and months when it begins to reveal colors. Winogradsky columns can become a rainbow of layers over time, each color a different type of bacteria. Here are a few you may encounter.

GREEN SULFUR BACTERIA

Plants are the best-known organisms that make use of photosynthesis, but some bacteria can also produce food this way. The color of green sulfur bacteria is made by chlorophyll, showing they are one of the few that can photosynthesize. Unlike plants, which release oxygen, green sulfur bacteria release, as the name indicates, sulfur. They are often found with other bacteria, notably those known as sulfate-reducing bacteria, which consume the sulfur waste from the green sulfur bacteria. In the microbial garden, the green sulfur may be accompanied by black sulfate-reducing bacteria.

SULFATE-REDUCING DELTA PROTEOBACTERIA

If you've ever stood over a marshy area only to be assaulted by the smell of rotten eggs, you've smelled sulfate-reducing bacteria. They're common and easy to detect with your nose. As the name implies, they consume sulfate. But they don't eat it; they breathe it in and expel hydrogen sulfide, the source of that stinky, rotten eggs odor. These black bacteria thrive anywhere with an abundant source of decaying organic matter. Interestingly, sulfate-reducing bacteria live in our intestines, and they are partly to thank for the unpleasant gaseous odors we often embarrassingly emit.

CYANOBACTERIA

Like green sulfur bacteria, cyanobacteria can also photosynthesize and may be easy to confuse with green sulfur in a Winogradsky column. Out in nature, they're incredibly abundant in many habitats, but they can be confused with algae and are largely impossible to differentiate without a microscope. They're not all green, though. Some can be orange, yellow, black, or brown. When cyanobacteria grow where algae is found, they often live on the periphery, but they can also be found mixed with the algae. However, cyanobacteria tends to grow in more extreme environments than algae, such as hot springs or glaciers, to avoid competition.

Learn the Language of Nature

SEASONS:
All

STUDY TOPIC:
Observation skills

oots standing in a line on the shore, a robin singing lustily from the top of a shrub, a squirrel clinging to a branch as still as a statue . . . these are all part of the language of nature. It's a complex language to be sure but, just like any foreign language, one that you can learn with immersion, time, and dedication. And when a naturalist begins to learn the language of nature, they start to understand that coots clustered on the shore instead of foraging in the water means an otter may be swimming nearby. The robin is singing as a threat to any other males who may dare intrude on his territory, and the squirrel is doing its best impression of a tree branch to hide from a raptor flying overhead.

Like all languages, there are no shortcuts to learning the language of nature, but there

The language of nature is complex and interconnected.

are simple tools and methods a naturalist can employ to learn in ways that are immersive and fun. The language of nature is multilayered. The easiest sounds to hear are from animals, like birdsong, calls, and alarms, which are well documented and simple to find samples of online and in apps. Naturalists may also be familiar with other sounds like crickets singing at night, a bear growling a warning, or a chipmunk chirping in alarm. But there is much more to the language of nature than animal sounds. It also includes the seasonal changes and the longer changes from year to year. It's the sound of the wind fluttering poplar leaves or the slow unfurling of fern fronds in the spring. It's recognizing the life cycle of wild organisms, what the clouds in the sky portend, and the smell of snow before it falls.

It's entirely possible for naturalists to learn to read the patterns and rhythms of the natural world if they start with the baseline. That means learning what is normal throughout the day and throughout the year in a particular place. Recognizing when certain flowers bloom each year, when the mayflies dance along the lake's edge, when the fireflies come out at night, when the owls head to their roost, the behavior of the thrush as it forages on the ground for food, the water level of the river in spring—when naturalists notice these throughout the year, they begin to recognize the rhythm of nature in a certain location or region. Then, naturalists can notice what's different when something is amiss and investigate to find out why those coots are standing on the lakeshore instead of frantically foraging in a cluster in the water—or, more subtly, why a plant is not blooming at the expected time, perhaps a sign of a fungal infection, invertebrates in the buds, or other mysterious causes.

Learning the baseline of the language of nature is perhaps one of the most relaxing and rewarding of a naturalist's activities because it's an excuse to spend time outside and do nothing, which we know is wonderful for our mental and physical health, as countless studies have shown. Becoming fluent in this language will also help hone your naturalist senses.

Sherlock Holmes once pointed out that seeing is more superficial, looking only at the surface, while observation is more complex and deeper. Seeing is walking through a forest and noticing there are tall trees and that it's sunny. Observing is smelling the resin of the pine trees warmed by the sun and hearing the crackling of dried pine needles underfoot and the nuthatch calling nearby. It's poking an orange mass on a log and feeling that it's spongy and damp. Observing is noting a pile of wood chips at the base of the tree and looking up to find a woodpecker's hole. Observing is noticing a rock with three species of lichens waging a very slow war over territory. Observing uses all of your senses all the time instead of simply seeing what's in front of you.

Two main methods to approach learning the language of nature are wandering and sitting. A popular concept is called a "sit spot," a single location that you visit as often as possible, where you simply sit and observe. Wandering keeps you moving, but unlike hiking, it is a slow and mindful walk. Either way, you have permission to just be outside and do nothing. There are no assignments, no homework. Simply observe with your senses. You might write notes, make sketches, or take photos and record sounds with a phone. Or you can simply immerse yourself without documenting anything except your mental experience. The key is immersion and consistency. Do it as often as possible, at a regular location, and at different times of the day and year.

Charles Darwin famously had a "thinking path" at Down House, his home in London. He wandered along this path every day, contemplating the wider mysteries of natural history while observing the nature of his own garden. He may have been best known for his observations on the Galápagos Islands, but his experiments in his home garden and regular wandering along his thinking path were instrumental in his work. As far roaming as many naturalists may be, most, if not all, have a local patch they are intimately familiar with.

Materials

Learning the language of nature requires no tools other than the senses you already possess. That said, there are a number of optional tools based on what approach you prefer to take.

Journal: The type of notebook is entirely up to you. It may be a sketchbook, a journal, waterproof paper, or a basic lined notebook. (See Project 1 on nature journaling for more information.)

Phone: Our phones are powerful tools for recording nature. If you have a smartphone, use it to make notes, record sounds and videos, and take photos, including macro, or close-up, shots with special lens attachments. Phones can also carry identification guides for everything from birds to nudibranchs and dragonflies. Just be sure to mute your phone while you're out, so you can focus on observing.

Binoculars: Because we don't have the vision of a hawk, binoculars can help aid our eyes and reveal distant birds or other things. Binoculars designed to also focus very close are great for looking at insects like butterflies and dragonflies that may be hard to approach for closer observation.

Hand Loupe: To see the opposite of things far away, a hand loupe can unveil the tiny world of lichens, mosses, invertebrates, and more. A magnifying glass works well, too, but a hand loupe is tiny, easy to carry, and powerful.

Outdoor Gear: The biggest barrier to most people's willingness to get and stay outside is discomfort (wet pants in the winter, for instance, are horribly uncomfortable). Getting the right outdoor gear can make all the difference. In summer, you may benefit from a sun hat, cooling sleeves, a neck gaiter, insect-resistant clothing, and water shoes that are suitable for walking. Winter items may include rechargeable hand warmers, a fleece-lined neck gaiter, waterproof and insulated boots, long underwear, a warm hat, gloves, and a warm coat.

Wet-weather items should absolutely include full raingear—not just water-resistant, but a fully waterproof jacket with a hood, pants, and shoes or boots, as well as a rain-proof cover for your bag. In all types of weather, consider carrying a waterproof seat pad in your field bag to keep your backside and/or knees dry and clean. You can make one from an old shower curtain, specialty cloth diaper fabric, tarp scrap, or other waterproof material—even a piece of cardboard covered with duct tape. It only needs to be compact and lightweight.

Steps

Choose a Location: The most important consideration when choosing your regular location for a sit spot or to wander is that it be practical. You're much more likely to develop a habit of going frequently if the location is easy to get to. It can be a local park, your own backyard, or even a vacant lot in the neighborhood. If you can't get out, it can be your own doorstep or even an open window of your house. There is no "right" location. If you realize you aren't going as much as you'd like to, find a more convenient spot.

Sit: If you choose to have a sit spot, settle in, get comfortable, and stay quiet. At first, it may seem like little is happening, but that's because simply by arriving in an area, you've disturbed the local wildlife. As you sit quietly, after about five to ten minutes, the birds and other animals will resume their activity, accepting you as part of the landscape. Then you can truly start to observe.

Wander: If you've chosen to wander, walk slowly without any real destination or objective in mind. If you're at a park with trails, take whatever path catches your attention. (Keep track of where you are though, so you don't get lost.) Stop frequently to listen and use your other senses. If you feel like it, try sitting for a while.

Record Observations: If you are keeping a nature journal and take it with you, write down your observations. Note the weather, time of day, and date. Make a sketch of the scene or take a photo. Take note of which flowers are blooming. Write down what birds you hear and see, and what insects you encounter. Make notes on their behavior. You don't have to know their names or identify them; it's fine to just describe them. If you want to try to identify them later, make as many notes as possible about their appearance and behavior. Write down anything else that comes to mind.

Ask Questions: Constantly ask questions about what you're observing, and write them down too. Why does the red dragonfly always return to the same perch, but the blue one never lands at all? Why does the grebe sometimes dive down into the water, but other times sink like a submarine? Why do those water beetles spin in endless circles?

Repeat: The key is to go out as often as possible, as many days a week as you can, even if it's just for a few minutes at a time. Start with short visits, and slowly build up the amount of time you spend sitting or wandering each time.

Change Your Focus: As you sit or wander, you may be drawn to the birds, trees, or flowers. The more obvious and bigger organisms will naturally draw your attention, but try to change your focus and look at something small, like the lichens on a rock or flies on a fungus. Consider the microscopic life in the moss growing on the side of a tree, and maybe even collect some to study under a microscope (see Project 6). Then step back and look at the overall setting. Where are you in the landscape and region? In a valley? On a mountain? What's the habitat? Forest? Wetland? Note the plant composition. Note the overall weather, and then notice how your location reflects the weather. Perhaps it's foggier than the surrounding hills, or the thick trees block the wind or provide shelter from the drizzle.

Challenge Yourself: At first, each visit may seem the same, but there is more going on than you can take in with your senses. Start to challenge yourself to find something you haven't noticed before. A jelly fungus growing on the side of a dead log, a water beetle coming to the surface of the pond every few minutes, a salamander hidden under a chunk of dead wood. It takes a lot of practice to "get your eyes in." Once you notice something, as is often the case, you'll start to see it everywhere and wonder how you never noticed it before. Even after years of observations, you will continue to discover things you never noticed before.

Also begin to push the boundaries of your senses. Take each in turn to focus on and stretch them further each day. Try to focus gently with your eyes and see what you notice in your peripheral vision. Close your eyes and try to note every single sound you can hear. Take deep breaths and see what scents you can detect, even if you can't identify them. Feel the breeze, mist, and sun on your skin, but also begin to make a habit of touching things you pass, whether that's a tree trunk, leaves, or feathers. Just remember to be cautious of poisonous plants or any with thorns.

If you can safely identify fruits, plants, and mushrooms, try a taste of something growing wild, like a salmonberry or blueberry, on your wander.

Be Patient: Just as for a foreign language, it takes times to learn the language of nature. Don't go out with expectations of what will happen or what you will see. You can't plan to see an owl dive down in front of you to catch a mouse, or a raccoon sprint across the path and up a tree to stare down at you. Every moment you spend outside is valuable and helps increase your understanding of the baseline of the language of the natural world.

Meditate and Be Mindful: If you practice meditation or mindfulness, try these practices outside. They are wonderful methods for observing and experiencing nature. With meditation, when you close your eyes and focus, you can start to hone your sense of hearing by noting all the different sounds, like birds singing or foraging in the dead leaves, a squirrel running up a tree, and the wind making two trees knock against each other. Similarly, mindfulness also helps train the senses by allowing you to be present in the moment and tuning in to the feeling of mist on your face or the scent of a blooming flower carried on the breeze.

Create a Curiosity Cabinet

SEASONS:
All

STUDY TOPICS:
Organization of natural history, managing a collection

Any number of things may cause a naturalist to stop in the midst of wandering, reach out, and pick it up to examine it more closely—for example, a dead bumble bee on the side of a path, a dragonfly wing under a spiderweb, a branch covered with lichens knocked down by the wind, or an empty turtle shell washed up on the lakeshore. Many of these items may make their way into a pocket or field bag, to be studied more closely under a microscope, drawn in detail, or otherwise observed and preserved. When a dead bumble bee is preserved in a clear vial or a turtle shell is set on a shelf, the naturalist is joining in the time-honored tradition of creating a curiosity cabinet—a valuable tool for learning about the natural world.

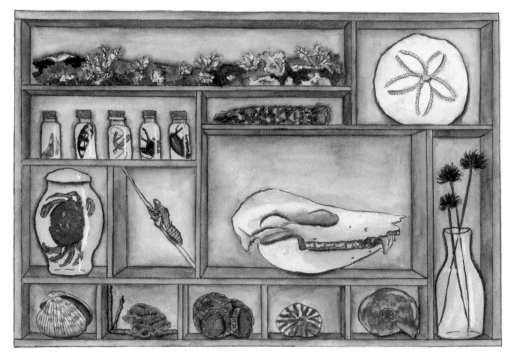

A simple curiosity cabinet

Curiosity cabinets may refer to a simple box, single shelf, whole cabinet, or even an entire room. Historically, they contained a wide variety of items, from shells, mounted butterflies, and feathers to giant taxidermized mammals. The only requirement was that the items sparked a sense of wonder and awe and inspired viewers to ask questions or stop and think.

These cabinets, also commonly known by the German name *Wunderkammer*, were popular between 1550 and around 1750. In the heyday of the curiosity cabinet, individuals from apothecaries to princes assembled cabinets—sometimes even multiple rooms—often referred to as theaters, and visiting one was an event. Cabinets were designed to be interactive, and the items inside were meant to be examined, pondered, and conversed over. Even the construction of some elaborate cabinets played into this, encouraging viewers to be interactive and discover hidden drawers and panels housing some of the rarest or most interesting objects, dramatically enhancing the awe and curiosity.

Scientific study was not the goal of those early cabinets, because this was the time before science broke and divided the world into smaller components in our unending quest to try and understand our messy, complex, and confusing cosmos. Instead, early collectors assembled pieces of the natural world as an attempt to make sense of it in a manageable size, organized with their own rationale and imaginations, often with spectacular results. They emphasized diversity, contrast, and abundance while focusing on the composition and aesthetics of the collection as a whole.

If you visit a naturalist today, you may spot remnants of those classic curiosity cabinets in their home, but the practice has largely been relegated to the past, like the study of natural history. However, there is an opportunity for modern naturalists to revisit the cabinets of wonder that inspired collectors, because even in a museum-dominated world, there is room to de-compartmentalize the natural world, and to use individual imagination and logic to make connections and see patterns in a way the rigid segregation and strict rules of museums and science often inhibit.

Although we know a lot more about the natural world than we did a few centuries ago, naturalists are still driven to make sense of what we encounter. A curiosity cabinet can help us organize our own thoughts and share them with others. The early alchemists and apothecaries created their collections for their professions, and their cabinets served a practical function as both laboratories to experiment and practice in and classrooms to teach apprentices. We, too, can echo the past and use cabinets as unofficial laboratories and classrooms.

Curiosity cabinets don't have to live in seclusion. Historically, cabinets were meant to be viewed by visitors, and some became so popular they turned into tourist destinations. The modern naturalist may not want to turn their home into a tourist stop, but creating and sharing your collection is a great opportunity to open conversations, exchange thoughts, and spark curiosity with like-minded people. A collection, presented in an inviting and interactive way, has the potential to not only instill wonder about the natural world in dinner guests but also start them on their own journey of becoming naturalists.

Naturalists have the freedom to arrange and design their cabinets in unique and creative ways. We can draw on the imaginative methods of old cabinets, in which items were arranged to tell a story or represent an idea. Collectors would put objects together with care and consideration to reveal secrets or illustrate connections between organisms and locations while paying close attention to the overall aesthetics. Sometimes they would use the cabinets to represent locations, either specific places or entire continents. Of course, naturalists can also choose to display their collection in the Linnean organization, which modern museums often use, and which groups items together by genus and family, showing a different sort of connection that is still steeped in history.

Another lesson that modern naturalists can take from curiosity cabinets of the past is that the displays included both natural and manmade or related items. Modern society largely draws a line between nature and human, but separating ourselves from nature is detrimental. We are part of nature, and nature is a huge part of us as well. Historical cabinets blurred that line in many ways, displaying a wonderful blend of natural history, the occult, magic, and folklore. "Mermaids," "unicorn horns," and "giant's bones" were displayed as some of the ultimate prizes. Today we know that such bones were from dinosaurs, the mermaids were cleverly constructed from monkeys and fish (objects of curiosity in their own right today), and unicorn horns were usually narwhal tusks.

The popularity of curiosity cabinets waned in the nineteenth century as science moved collections to museums dominated by the new Linnean and Darwinian ideas. Many curiosity cabinet collections served as the foundation of modern museums. At that time, museums were designed to teach, not explore, and there was a loss of the spiritual thinking and connection to magic and folklore. A few individuals since that time have sidestepped the rigid organization of science and museums and embraced a more idiosyncratic system.

In a world where not only natural history is forgotten but nature is largely overlooked by many people, naturalists are in a unique position to bring back the tradition of curiosity cabinets and share them to instill wonder and awe for our natural world with our friends and neighbors.

Materials

Cabinet: There is an incredible variety of possibilities for curiosity cabinets. It all depends on the size of your collection and aesthetic preference. You can start small with a simple box, and your cabinet can expand as your collection grows. You might start with something like a cigar box or a dedicated shelf on a bookcase. Thrift stores have a wealth of options, from old jewelry boxes to empty boxes from painting sets, mancala trays, wooden soda boxes, spice jar boxes, and wine crates. These potential cabinets can be refinished as well. For larger collections, you can use entire bookshelves or other types of cabinets.

If you're handy, you can construct your own cabinet or boxes from wood, using balsa wood as dividers to make a classic display similar to the much sought-after typesetter's tray.

Containers: Some items can be displayed directly on a shelf, but many nature objects are fragile or suscepti-

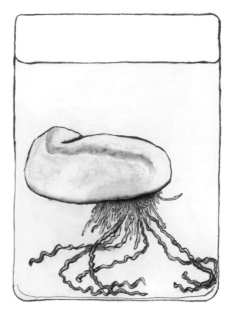

A preserved Portuguese man o' war

ble to pests and need to be protected. Again, a thrift store is a wonderful source for these containers. Spice jars and other kitchen containers make excellent storage options for lichens, fungi, wasp nests, shells, acorns, plant galls, and countless other items. Decorative jars, bowls, and other clear containers are abundant at thrift stores and can hold all manner of nature objects. Vases can hold interesting branches, dried flowers, grasses, and other taller objects. Tiny craft jars with cork lids can protect beetles, flies, or other insects found dead on the ground. A few invertebrates, like spiders and worms, need to be preserved in alcohol and require special sealed vials (ones that won't leak and will prevent air from evaporating the alcohol) that can be found at entomological supply stores online.

Alcohol: For the most part, invertebrates that need to be preserved wet can be kept in standard rubbing alcohol that is at least 70 percent ethanol. Some invertebrates do better with 95 percent ethanol, but for a naturalist's personal collection, standard rubbing alcohol should be sufficient.

Moisture Absorbers: Depending on where you live and what's in your collection, you could put some type of moisture absorber in your cabinets to prevent anything from getting moldy or otherwise ruined by moisture or humidity. Hang on to those little silica packs that come in shoes or vitamins, or buy special moisture absorbers and hide them away in your cabinets.

Steps

Collect: This is a lifelong, ongoing process. Any naturalist will forever have rocks, acorns, shells, and any number of other objects in their pockets or bags. There are no guidelines for what to collect for a curiosity cabinet. If you find it interesting and it's legal and ethical to collect, add it to your collection.

Naturalists may choose to collect whatever interests them, but there are other methods as well. When you go on a vacation, for example, you could collect some notable things from that area, like a leaf, a seed pod, lichens, or moss, or whatever else makes that place unique. Perhaps you want to document the nature of your yard or local nature spot. Or maybe you want to focus your theme on a subject instead. For example, you could organize a curiosity cabinet based on a tree, focusing on a certain species, which may include bark, stems, flowers, and seeds; the lichen and mosses that grow on the plant; the insects that eat it; and so on.

Note

Before you collect anything, you must be responsible for knowing the rules and laws governing a particular location. Never take anything from private land without permission, and note local laws for parks and other nature areas. National parks in the United States prohibit visitors from collecting anything at all. It is almost universally illegal to possess certain objects, like eggs, so research your county, state, or other laws you are subject to.

In historical cabinets, if an object couldn't be acquired, a painting was often used in its place. If you want to include something that's unethical or illegal to collect or impossible to find, consider adding a photo, drawing, model, or some other representation of that object instead. Consider also how your actions may affect the local natural world.

Clean and Prepare: Examine the object before you put it into your collection. It may need to be washed, especially if it's from an animal and may start to rot or stink. You can research methods for cleaning bones and other items online. If it's an object that may harbor invertebrates, especially any that may damage your collection, freeze it for a couple of days to kill any pests.

Preserve: Many natural objects are ephemeral and meant to decompose back into the ground, so they need to be taken care of to last in a curiosity cabinet. Keep fragile items like wasp nests, butterfly wings, and dragonfly exuviae in jars where they won't crumble if bumped. Protect your invertebrates, feathers, and animal items from carpet beetles and other pests with an airtight lid. Carpet beetle larvae can get into containers with the smallest opening.

Warning

Carpet beetles and other pests can wreak havoc on a collection. Carpet beetle larvae will eat anything from animal parts, including the glue in vintage books. They can destroy your found insects among other items. The adults look like tiny lady beetles with a light and dark brown pattern. Their brown, hairy larvae are not often seen, but their skin molts are.

Another sign is a pile of fine, powdery dust on the shelf under objects. Survey your cabinets regularly for signs of the beetles or other pests. You may need to freeze infected objects for a few days. Vacuum or sweep the room your cabinet is in often to keep your collection safe from potential pests.

Record: If you choose, make a note on each object or put a label on the container with relevant information. Printer labels work well for sticking inside jar lids or on the bottom of containers with brief notes. If you follow in the tradition of letting visitors view your curiosity cabinet, adding decorative labels will help explain what certain items are.

Organize: Decide on an organizational structure for your cabinet as a whole. How do you wish to present your items? You may choose arbitrary arrangements or present your objects as a story. Following the organization of museums and grouping things according to genus or family is always an option. Your collection may be best considered by habitat type, location, or season or based on species interactions. If you choose an unorthodox arrangement with your own idiosyncratic methods, perhaps you'll uncover your own secrets of the natural world.

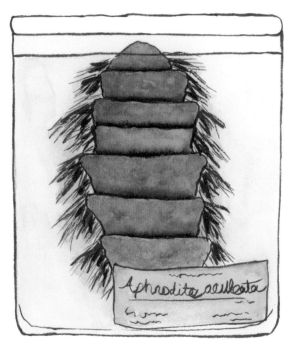

A preserved sea mouse

If you're looking for inspiration, research historical curiosity cabinets, look at modern versions, or visit museums to get ideas about how to arrange displays. Imagination and creative thinking can result in some interesting and unique displays, which may lead to new ideas or revelations.

Display Ideas: How do you want to display your collections? You might create a diorama in a box or jar to represent the ground of an oak woodland and include acorns and oak leaves, a shed snake skin, a woodpecker skull (where legal), and a vial of filbert weevils. Another, similar idea is to make a location jar, which is a selection of objects from a certain location. For example, from the beach, put sand in the bottom, a few shells, dried-up eggs from whelks or a skate's egg case, washed-up fish bones, and a few bits of dried-up seaweed. Or take a page from the classic cabinets and merge folklore and nature like contemporary artist Françoise Pétrovitch did when she created a cabinet containing a classic wolf painting, a wolf skull, and Little Red Riding Hood's shoes.

Acknowledgments

The Naturalist at Home arose from a series of articles I wrote on my website after the publication of my first book, *Nature Obscura*. Many people had asked me how to do specific things I had mentioned, like how to find tardigrades, so I began writing instructions, which were well received. I am eternally grateful that Mountaineers Books editor in chief Kate Rogers recognized not only the potential of that material but also just how perfect a follow-up to *Nature Obscura* this book would be. It was too good an idea to pass by, and researching and writing it took me on a new journey of learning ever more about the study of natural history as well as spending more time creating art.

This project would not have been possible without my wonderful patrons on Patreon whose ongoing support helps fund my website, allowing me a place to explore new ideas in natural history–related writing. The articles I have written on my website helped lead to this book and other opportunities. I appreciate each and every one of my patrons, past and present.

My gratitude goes to editor Beth Jusino, whose insight and thoughtful input as a new naturalist was incredibly valuable. I'm also appreciative of the rest of the Mountaineers Books team who provided their expertise and help, especially project editor Laura Shauger, copy editor Jennifer Kepler, designer Jen Grable, and proofreader Joeth Zucco. I'm grateful to know that I inspired them along the way with the projects in this book.

Writing is usually a solitary existence and doubly so during a global pandemic. Although I like spending time alone, I am grateful for writer friends I commiserated with and connected with virtually, including Sarah Swanson and Eric Wagner. I'm also grateful for the ongoing support of my family. My daughter, Amalie, has been the most supportive of all, and for this book she was always full of praise for my art and rewarded me with "Zelda time" and treats after I finished each chapter.

Resources

There is an abundance of resources, both modern and vintage, to fill the curious naturalist's library shelves. The resources listed below are a few on my shelves or resources I have found useful in my naturalist journey. Look for vintage books online where you can find scanned, free copies to download and read.

General Resources

Barber, Lynn. *The Heyday of Natural History.* Garden City, NY: Doubleday, 1980.

Biodiversity Heritage Library. Headquartered at the Smithsonian Libraries and Archives, Washington, DC. www.biodiversitylibrary.org.

Chick, Andrew. *Insect Microscopy.* Ramsbury, Wiltshire: Crowood Press, 2016.

Comstock, Anna Botsford. *Handbook of Nature-Study.* Ithaca, NY: Comstock, 1967.

Durrell, Gerald. *The Amateur Naturalist.* New York: Alfred A. Knopf, 1988.

Gerlach, John. *Digital Nature Photography: The Art and the Science.* Burlington, MA: Focal Press, 2015.

Gooley, Tristan. *The Lost Art of Reading Nature's Signs: Use Outdoor Clues to Find Your Way, Predict the Weather, Locate Water, Track Animals—and Other Forgotten Skills.* New York: Experiment, 2015.

Headstrom, Richard. *Adventures with a Hand Lens.* New York: Dover, 1976.

Leslie, Clare Walker. *The Curious Nature Guide: Explore the Natural Wonders All Around You.* North Adams, MA: Storey, 2015.

Macmillian. *The Way Nature Works.* New York: Macmillian, 1992.

Moss, Stephen. *The Bumper Book of Nature: A User's Guide to the Outdoors.* New York: Harmony Books, 2010.

Books About and by Naturalists

Attenborough, David. *Life on Air: Memoirs of a Broadcaster.* Princeton, NJ: Princeton University Press, 2002.

Blackburn, Julia. *Charles Waterton, 1782–1865: Traveller and Conservationist.* London: Vintage, 1997.

Bonta, Marcia Myers. *Women in the Field: America's Pioneering Women Naturalists.* College Station: Texas A&M University Press, 1991.

Brenner, Kelly. *Nature Obscura: A City's Hidden Natural World.* Seattle: Mountaineers Books, 2020.

Carson, Rachel. *The Edge of the Sea.* Boston: Houghton Mifflin, 1955.

Crowder, William. *A Naturalist at the Seashore.* New York: Century, 1928.

Gooley, Tristan. *The Natural Navigator: The Rediscovered Art of Letting Nature Be Your Guide.* New York: Experiment, 2012.

Haupt, Lyanda Lynn. *Crow Planet: Essential Wisdom from the Urban Wilderness.* New York: Little, Brown, 2009.

———. *The Urban Bestiary: Encountering the Everyday Wild.* New York: Little, Brown, 2013.

Holden, Edith. *The Country Diary of an Edwardian Lady.* New York: Rizzoli, 2018.

Keffer, Ken. *Earth Almanac: Nature's Calendar for Year-Round Discovery.* Seattle: Skipstone, 2020.

Mabey, Richard. *The Unofficial Countryside.* Toller Fratrum, Dorset: Little Toller Books, 2010.

McAnulty, Dara. *Diary of a Young Naturalist.* Dorset: Little Toller Books, 2020.

Nisbet, Jack. *The Collector: David Douglas and the Natural History of the Northwest.* Seattle: Sasquatch Books, 2009.

Packham, Chris. *Fingers in the Sparkle Jar: A Memoir.* London: Ebury Press, 2016.

Preston, Diana, and Michael Preston. *A Pirate of Exquisite Mind: Explorer, Naturalist, and Buccaneer: The Life of William Dampier.* New York: Walker, 2004.

Shepherd, Nan. *The Living Mountain: A Celebration of the Cairngorm Mountains of Scotland.* Edinburgh: Canongate, 2008.

Stott, Rebecca. *Darwin and the Barnacle: The Story of One Tiny Creature and History's Most Spectacular Scientific Breakthrough.* New York: W. W. Norton, 2004.

Tinbergen, Niko. *Curious Naturalists.* Amherst: University of Massachusetts Press, 1984.

Todd, Kim. *Chrysalis: Maria Sibylla Merian and the Secrets of Metamorphosis.* Boston: Mariner Books, 2007.

Wilson, Edward O. *Naturalist.* Washington, DC: Island Press, 1994.

Wulf, Andrea. *The Invention of Nature: The Adventures of Alexander von Humboldt, the Lost Hero of Science.* London: John Murray, 2015.

Community Science Projects

Community science projects (and websites) can be ephemeral. If one that catches your attention has ended, search online and you may find something similar. Hundreds of local and regional community science programs are happening all the time. The ones listed below are some of the longer-running projects or directories to projects.

» Audubon Christmas Bird Count: www.audubon.org/conservation/science/christmas-bird-count
» Budburst: https://budburst.org
» Bumble Bee Watch: www.bumblebee watch.org
» Chronolog: www.chronolog.io
» CitSci: www.citsci.org
» eBird: https://ebird.org
» Great Backyard Bird Count: www.birdcount.org
» iNaturalist: www.inaturalist.org
» National Moth Week: https://national mothweek.org
» National Phenology Network: www.usanpn.org
» NestWatch: https://nestwatch.org
» SciStarter: https://scistarter.org
» Western Monarch Count: www.western monarchcount.org
» Zooniverse: www.zooniverse.org

Nature Journaling

Brown, Jo. *Secrets of a Devon Wood: My Nature Journal.* London: Short Books, 2020.

Haizlett, Rosalie. *Watercolor in Nature: Paint Woodland Wildlife and Botanicals with 20 Beginner-Friendly Projects.* Salem, MA: Page Street, 2021.

Laws, John Muir. *The Laws Guide to Nature Drawing and Journaling.* In collaboration with Emilie Lygren. Berkeley, CA: Heyday, 2016.

Leslie, Clare Walker, and Charles E. Roth. *Keeping a Nature Journal: Discover a Whole New Way of Seeing the World Around You.* North Adams, MA: Storey, 2003.

Tomlinson, Susan Leigh. *How to Keep a Naturalist's Notebook.* Mechanicsburg, PA: Stackpole Books, 2009.

Pond Dipping and Wetland in a Jar

Hernick, Linda VanAller. *Most Wonderful in the Smallest: A Year in Pursuit of Common Freshwater Microorganisms.* Newark, OH: McDonald & Woodward, 2017.

Klots, Elsie B. *The New Field Book of Freshwater Life.* New York: Putnam, 1966.

Moss, Brian. *Ponds and Small Lakes: Microorganisms and Freshwater Ecology.* Exeter, UK: Pelagic, 2017.

Reid, George K. *Pond Life: A Guide to Common Plants and Animals of North American Ponds and Lakes.* New York: St. Martin's Press, 2001.

Thorp, James H., and Christopher Rogers. *Field Guide to Freshwater Invertebrates of North America.* Burlington, MA: Academic Press, 2011.

Voshell, J. Reese. *A Guide to Common Freshwater Invertebrates of North America.* Illustrated by Amy Bartlett Wright. Blacksburg, VA: McDonald & Woodward, 2002.

Waldbauer, Gilbert. *A Walk Around the Pond: Insects In and Over the Water.* Cambridge, MA: Harvard University Press, 2006.

Walstad, Diana. *Ecology of the Planted Aquarium: A Practical Manual and Scientific Treatise for the Home Aquarist.* Chapel Hill, NC: Echinodorus, 2012.

Animal Signs

Brown Jr., Tom. *Tom Brown's Field Guide to Nature Observation and Tracking.* With Brandt Morgan. Illustrated by Heather Bolyn. New York: Berkley Books, 1983.

Eiseman, Charley, and Noah Charney. *Tracks and Sign of Insects and Other Invertebrates: A Guide to North American Species.* With John Carlson. Harrisburg, PA: Stackpole Books, 2010.

Elbroch, Mark. *Mammal Tracks and Sign: A Guide to North American Species.* Mechanicsburg, PA: Stackpole Books, 2003.

Murie, Olaus J. *A Field Guide to Animal Tracks.* Boston: Houghton Mifflin, 1954.

Sheldon, Ian. *Animal Tracks of Washington and Oregon.* Redmond, WA: Lone Pine, 1997.

Stokes, Donald, and Lillian Stokes. *A Guide to Animal Tracking and Behavior.* Boston: Little, Brown, 1986.

Tardigrades and Moss Animals

Glime, Janice M. *Bryophyte Ecology.* Ebook sponsored by Michigan Technological University and the International Association of Bryologists. Last updated January 23, 2022. https://digitalcommons.mtu.edu/bryophyte-ecology.

Kimmerer, Robin Wall. *Gathering Moss: A Natural and Cultural History of Mosses.* Corvallis: Oregon State University Press, 2003.

Schofield, W. B. *Some Common Mosses of British Columbia.* Victoria: Royal British Columbia Museum, 1992.

Moths

Gandy, Matthew. *Moth.* London: Reaktion Books, 2016.

Himmelman, John. *Discovering Moths: Nighttime Jewels in Your Own Backyard.* Camden, ME: Down East Books, 2002.

Powell, Jerry, and Paul Opler. *Moths of Western North America.* Berkeley: University of California Press, 2009.

General Invertebrates

Berenbaum, May R. *Bugs in the System: Insects and Their Impact on Human Affairs.* Reading, MA: Helix Books, 1995.

Buchsbaum, Ralph. *Animals Without Backbones.* Harmondsworth, England: Penguin Books, 1951.

BugGuide (website). www.bugguide.net.

Evans, Howard Ensign. *Life on a Little-Known Planet: A Biologist's View of Insects and Their World.* New York: Lyons & Burford, 1993.

Hubbell, Sue. *Broadsides from the Other Orders: A Book of Bugs.* Illustrated by Dimitry Schidlovsky. New York: Random House, 1993.

Johnson, Elizabeth A., and Kefyn M. Catley. *Life in the Leaf Litter.* New York: American Museum of Natural History, 2002. www.amnh.org/content/download/35188/518925/file/LifeInTheLeafLitter.pdf.

Murray, Andy. *A Chaos of Delight.* www.chaosofdelight.org.

Nardi, James B. *Life in the Soil: A Guide for Naturalists and Gardeners.* Chicago: University of Chicago Press, 2009.

Sverdrup-Thygeson, Anne. *Extraordinary Insects: Weird, Wonderful, Indispensable, the Ones Who Run Our World.* London: Mudlark, 2019.

Wilson, Joseph, and Olivia Messinger Carril. *The Bees in Your Backyard: A Guide to North America's Bees.* Princeton, NJ: Princeton University Press, 2015.

Habitat Terrariums

Humphreys, Henry Noel. *The Butterfly Vivarium; or, Insect Home: Being an Account of a New Method of Observing the Curious Metamorphoses of Some of the Most Beautiful of Our Native Insects. Comprising also a Popular Description of the Habits and Instincts of Many of the Insects of the Various Classes Referred to; with Suggestions for the Successful Study of Entomology by*

Means of an Insect Vivarium. London: W. Lay, 1858.

Ward, Nathaniel Bagshaw. *On the Growth of Plants in Closely Glazed Cases.* London: J. Van Voorst, 1852.

Spiderwebs

Adams, R. J. *Field Guide to the Spiders of California and the Pacific Coast States.* Berkeley: University of California Press, 2014.

Brunetta, Leslie, and Catherine L. Craig. *Spider Silk: Evolution and 400 Million Years of Spinning, Waiting, Snagging, and Mating.* New Haven, CT: Yale University Press, 2010.

Crompton, John. *The Spider.* New York: Nick Lyons Books, 1987.

Hillyard, Paul. *The Book of the Spider: From Arachnophobia to the Love of Spiders.* London: Hutchinson, 1994.

Mushroom Spores

Arora, David. *Mushrooms Demystified.* Berkeley, CA: Ten Speed Press, 1986.

Stamets, Paul. *Mycelium Running: How Mushrooms Can Help Save the World.* Berkeley, CA: Ten Speed Press, 2011.

Stephenson, Steven L. *The Kingdom Fungi: The Biology of Mushrooms, Molds, and Lichens.* Portland, OR: Timber Press, 2010.

Winkler, Daniel. *Fruits of the Forest: A Field Guide to Pacific Northwest Edible Mushrooms.* Seattle: Mountaineers Books, 2022.

Slime Molds

Lister, Arthur, and Gulielma Lister. *A Monograph of the Mycetozoa: A Descriptive Catalogue of the Species in the Herbarium of the British Museum.* London: Trustees of the British Museum, 1911.

Lister, Gulielma. *The Mycetozoa: A Short History of Their Study in Britain; an Account of Their Habitats Generally; and a List of Species Recorded from Essex.* Stratford: Essex Field Club, 1918.

Lloyd, Sarah. *Where the Slime Mould Creeps: The Fascinating World of Myxomycetes.* Birralee, Tasmania: Tympanocryptis Press, 2014.

Niblett, Carrie. *The Curious Observer's Guide to Slime Mold of UC Santa Cruz and Beyond.* Santa Cruz: UCSC Natural Reserves, 2017.

Sharp, Jasper. *The Creeping Garden: Irrational Encounters with Plasmodial Slime Moulds.* Godalming, Surrey: Alchimia, 2015.

Stephenson, Steven L. *Myxomycetes: A Handbook of Slime Molds.* Portland, OR: Timber Press, 1994.

Tree Bark

Capon, Brian. *Botany for Gardeners.* Portland, OR: Timber Press, 2010.

Pollet, Cedric. *Bark: An Intimate Look at the World's Trees.* London: Frances Lincoln, 2010.

Sibley, David Allen. *The Sibley Guide to Trees.* New York: Knopf, 2009.

Vaucher, Hugues. *Tree Bark: A Color Guide.* Portland, OR: Timber Press, 2010.

Plant Pressing

DiNoto, Andrea, and David Winter. *The Pressed Plant.* New York: Harry N. Abrams, 1999.

MacFarlane, Ruth B. (Alford). *Collecting and Preserving Plants.* New York: Dover, 1994.

Microbes

Dyer, Betsey Dexter. *A Field Guide to Bacteria.* Ithaca, NY: Cornell University Press, 2003.

The Language of Nature

Gooley, Tristan. *How to Read Nature: Awaken Your Senses to the Outdoors You've Never Noticed.* New York. Experiment, 2017.

Williams, Florence. *The Nature Fix: Why Nature Makes Us Happier, Healthier, and More Creative.* New York. W. W. Norton, 2017.

Young, Jon. *What the Robin Knows: How Birds Reveal the Secrets of the Natural World.* Boston: Mariner Books, 2017.

Curiosity Cabinets

Davenne, Christine. *Cabinets of Wonder: A Passion for Collecting.* New York: Abrams, 2012.

Grice, Gordon. *Cabinet of Curiosities: Collecting and Understanding the Wonders of the Natural World.* New York: Workman, 2015.

Mauriès, Patrick. *Cabinets of Curiosities.* London: Thames & Hudson, 2019.

About the Author

A naturalist, writer, and artist based in Seattle, **Kelly Brenner** is the author of *Nature Obscura: A City's Hidden Natural World*, a finalist for the PNBA and Washington State book awards. Her writing has appeared in *Popular Science, Crosscut, National Wildlife Magazine,* and *The Open Notebook,* among others.

Brenner earned a bachelor's degree in landscape architecture from the University of Oregon and a certificate in nonfiction writing from the University of Washington. She shares her observations of the natural world, as well as folklore, books, and poetry, on metrofieldguide.com.

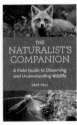